Reflecting Warfighter Needs in Air Force Programs

Prototype Analysis

Paul K. Davis, Richard Hillestad, Duncan Long, Paul Dreyer, Brandon Dues

Prepared for the United States Air Force

PROJECT AIR FORCE

The research described in this report was sponsored by the United States Air Force under Contract FA7014-06-C-0001. Further information may be obtained from the Strategic Planning Division, Directorate of Plans, Hq USAF.

Library of Congress Cataloging-in-Publication Data

Reflecting warfighter needs in Air Force programs : prototype analysis / Paul K. Davis ... [et al.].
    p. cm.
  Includes bibliographical references.
  ISBN 978-0-8330-4949-0 (pbk. : alk. paper)
  1. United States. Air Force—Management. 2. United States. Air Force—Planning. 3. United States. Air Force—Combat sustainability. 4. United States. Air Force—Operational readiness. 5. Close air support.
I. Davis, Paul K., 1943-

  UG633.R3925 2010
  358.4'168340973—dc22

                                                                              2010029614

The RAND Corporation is a nonprofit research organization providing objective analysis and effective solutions that address the challenges facing the public and private sectors around the world. RAND's publications do not necessarily reflect the opinions of its research clients and sponsors.

RAND® is a registered trademark.

Published 2010 by the RAND Corporation
1776 Main Street, P.O. Box 2138, Santa Monica, CA 90407-2138
1200 South Hayes Street, Arlington, VA 22202-5050
4570 Fifth Avenue, Suite 600, Pittsburgh, PA 15213-2665
RAND URL: http://www.rand.org/
To order RAND documents or to obtain additional information, contact
Distribution Services: Telephone: (310) 451-7002;
Fax: (310) 451-6915; Email: order@rand.org

Preface

This technical report describes the first phase of a Project AIR FORCE study, "Measuring Combat Effectiveness and Integrating Effectiveness Data into the Air Force Corporate Process," conducted in 2008 for the Office of Warfighting Integration and Chief Information Officer (SAF/XC). The study was requested by the Air Force with the purpose of finding ways to improve the ability to reflect warfighter needs in Air Force programs. The primary conclusions were briefed in June 2008 to Lt Gen Michael W. Peterson, who was then SAF/XC; the current SAF/XC is Lt Gen William Lord. The research reported here was conducted within the Force Modernization and Employment Program of RAND Project AIR FORCE.

Related RAND Corporation documents include the following:

- *Enhancing the Integration of Special Operations and Conventional Air Operations: Focus on the Air-Surface Interface*, by Jody Jacobs, Gary McLeod, and Eric V. Larson, Not Available to the General Public.
- *Technologies and Tactics for Improved Air-Ground Effectiveness*, by Jody Jacobs, Leland Joe, David Vaughan, Diana Dunham-Scott, Lewis Jamison, and Michael Webber, Not Available to the General Public.

The report will be useful to those who are responsible for balancing Air Force requirements and programs, including those that cross traditional "stovepipes."

RAND Project AIR FORCE

RAND Project AIR FORCE (PAF), a division of the RAND Corporation, is the U.S. Air Force's federally funded research and development center for studies and analyses. PAF provides the Air Force with independent analyses of policy alternatives affecting the development, employment, combat readiness, and support of current and future aerospace forces. Research is conducted in four programs: Force Modernization and Employment; Manpower, Personnel, and Training; Resource Management; and Strategy and Doctrine.

Additional information about PAF is available on our Web site:
http://www.rand.org/paf

Contents

Figures

Tables

Summary

Objective

This report documents a phase-one effort to develop new methods for the Air Force to use in ensuring that warfighter needs are adequately represented as the Air Force puts together its program and budget for the Department of Defense's (DoD's) Planning, Programming, Budgeting, and Execution process. We designed generic methods and then illustrated them for a single, concrete mission area: close air support (CAS). Our prototype analysis with notional data was concrete enough to demonstrate the primary concepts and analytic methods. The intended next stage of this research will flesh out the mission area with better data, apply the approach to additional mission areas, address trade-offs that cross mission areas (an especially difficult issue), and examine how the methods might be used within the current program-building process, or how that process might be modified.

Approach

When requesting our study, the Air Force expressed concern that, in its current system with current analytical methods, headquarters-level program decisions sometimes fail to give sufficient priority to requests important to meeting warfighter needs. Rather than starting with a review of current Air Force methods and processes, RAND was asked to take an independent cut at the challenge of improving the characterization and communication of warfighter needs for use in program building. The intent was to use analytical methods that would not only be sound but that would also translate readily into a narrative that could be communicated effectively to reviewers at all stages of the Air Force program-building process. That narrative should have an operational flavor clearly relevant to warfighting.

Our generic approach for work within a given mission area is summarized in Figure S.1. It is an adaptation of a methodology developed for the Under Secretary of Defense for Acquisition, Technology, and Logistics (USD/AT&L). Step One is to sharpen understanding of the mission area and its seams with other mission areas (where many problems reside). Step Two develops a sense of the scenario space (i.e., the full range of potential operational contexts) and then defines a small "spanning set" of test cases that, taken together, allow stressful tests.[1] If a programmed capability for the mission area does well by all such tests, the force will probably be able to deal effectively with actual situations as they arise. If limited resources dictate a program that still has shortfalls, the shortfalls can at least be well understood in warfighter-meaningful terms. (See pp. 3–4.)

Figure S.1
Schematic of Analytic Process

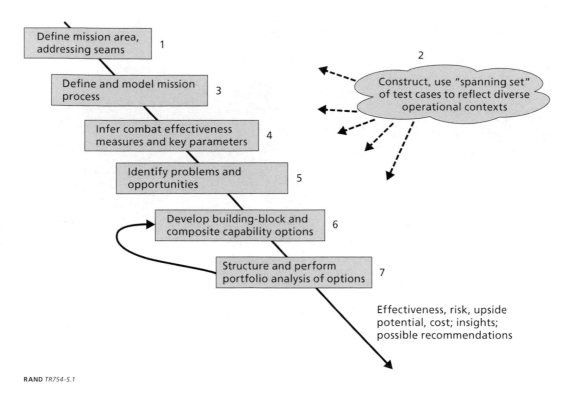

RAND *TR754-S.1*

Given a spanning set of test cases, the methodology calls for descriptions of the operational processes required to conduct the mission. This involves (Step Three) a "capability model," i.e., a relatively high-level, transparent, and parametric model suitable for uncertainty analysis in the program-development context. Such a model allows (Step Four) identification of *natural* measures of combat effectiveness—i.e., measures significant to warfighters and capability developers who think in operational terms. Such measures are often very different from and better than the measures and metrics conceived a priori. (See pp. 13–21.)

By examining current capabilities, it is possible to identify (Step Five) both shortcomings and potential opportunities. Doing so combines with other considerations, such as technology opportunities, to construct (Step Six) potential improvement options—both "building-block options" that address individual aspects of the mission-capability system and "composite options" that combine building-block options sensibly—so as to improve overall system performance and package matters suitably for program analysis. At this point, the composite options are evaluated (Step Seven) in a portfolio-analysis structure that highlights criteria for assessing warfighting value. The criteria relate to different classes of operation, such as supporting large-scale friendly maneuvers versus supporting lengthy but less intense stabilization operations. Subordinate criteria distinguish among results for different environments and threat capabilities. (See pp. 23–29.)

Portfolio Analysis to Assist in Resource Allocation

Program building in DoD and the military departments is essentially a process of portfolio management (pp. 31–39). The program should ensure a diversity of capabilities suitable for dealing with a wide range of potential challenges, with suitability judged by a number of very different criteria. The result is a need for "balancing" efforts and investments. Recently, this has been reflected by the Secretary of Defense asking the military departments to rebalance their portfolios to better reflect a world in which irregular warfare is now the norm, but maintaining and improving capabilities for high-end war remains necessary (Gates, 2009).

Such balancing is facilitated by appropriate analytical tools. In our approach, we use the RAND-developed Portfolio Analysis Tool (PAT) (Davis and Dreyer, 2009) and a suitable model to generate PAT's data. The portfolio analysis can be summarized in familiar stoplight charts, but with several additional features. After seeing a high-level assessment, one can drill down or "zoom" into detail so as to better understand the criteria that have been used, including assumptions about the threat environment. Further, one can zoom to a level of detail that shows *why*, in operational terms, a given option is better than another. One can visualize how much an option's performance is hampered by poor communications, inadequate situational awareness of where friendly forces are operating, inadequate responsiveness due to sluggish command and control (C2), or inadequate human resources (including forward observers). Such scorecard zooming (see Figure S.2) can often provide a visual audit trail, as indicated by the arrows, for illustrative drilldowns. In Figure S.2, the first column of the top scorecard is the result of analysis at the second level of detail, and the second column of the middle scorecard is the result of analysis at the third level.

Decisionmakers are shown options at the scorecard level, where they see multicriteria evaluations rather than evaluations in terms of some net effectiveness. For the illustrative analysis summarized in Figure S.2, the criteria include the options' effectiveness in each of four classes of CAS operations (i.e., four "scenarios" that constitute an approximate spanning set), their value for "other" (non-CAS) missions, and their risks. In the figure, the planning set consists of Maneuver A, Maneuver B, Stabilization A, and Stabilization B, and the two right-most columns indicate the "other" missions and risks.

After decisionmakers have gained the insights from this multicriteria evaluation, the final element of the portfolio analysis is to compare the options not only by effectiveness but also by cost. Unlike traditional cost-effectiveness calculations, RAND's methodology recognizes that the simplifying calculations of overall (i.e., net) "effectiveness" depend crucially not only on assumptions but also on judgments regarding the relative significance of such criteria as minimizing collateral damage on the one hand and maximizing support to maneuver forces on the other. Figure S.3 illustrates one form of results: It shows a *cost-effectiveness landscape* rather than a cost-effectiveness ratio at some specific cost level. Further, it shows how the landscapes vary across strategic *perspectives* that combine results across criteria in different ways. For example, in Figure S.3, the option entitled "Large JTAC and C2"[2] looks much more cost-effective if decisionmakers focus on the stabilization challenge (right panel) than if they focus on support-of-maneuver operations (left panel). Other perspectives can and should be considered as discussed in the main text.

Figure S.2
Portfolio Analysis of Alternatives, with Explanations Provided by Zooming (notional data)

Measures	Maneuver A	Maneuver B	Stabiliz. A	Stabiliz. B	Other Missions	Risks
	Detail	Detail	Detail	Detail	Detail	Detail
Investment Options	1	1	1	1	1	0
Baseline	R	R			O	G
JTAC Package + C2	Y	Y	O	O	O	LG
Large JTAC and C2 Package	Y	Y	Y	Y	O	LG
Sit. Aware. + C2 Package	Y	Y	LG	Y	LG	Y
Kill Package + C2	O	O	O	O	Y	Y

Level 1 Measure	Maneuver A					
Level 2 Measure	Perc. of Targets Killed	Time to Kill Target Score	Likelihood of Bad Effects Score	Efficiency	Warning	
Investment Option						Maneuver A Score
Baseline	O	R	Y	O		R
JTAC Package + C2	LG	Y	LG	G		Y
Large JTAC and C2 Package	LG	Y	LG	G		Y
Sit. Aware. + C2 Package	G	Y	G	G		Y
Kill Package + C2	Y	O	LG	Y		O

Level 2 Measure	Time to Kill Target Score						
Scoring Method	Thresholds						
Level 3 Measure	Detection Delay	Assignment Delay	Transit Time to Target	Attack Time	Prob (First-Time Kill)	Calc. Time to Kill Target	
Investment Option							Target Score Score
Baseline	LG	O	O	R	R	R	R
JTAC Package + C2	LG	LG	O	Y	G	Y	Y
Large JTAC and C2 Package	LG	LG	O	LG	G	Y	Y
Sit. Aware. + C2 Package	G	LG	O	Y	G	Y	Y
Kill Package + C2	LG	LG	O	O	O	O	O

Figure S.3
Cost-Effectiveness Curves as a Function of Perspective (notional data)

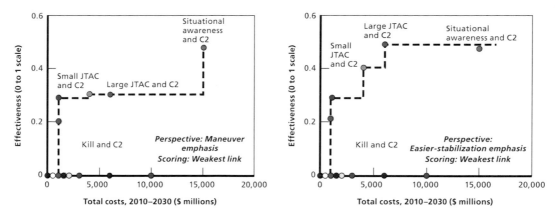

NOTE: Charts plot effectiveness versus cost for different perspectives. Each point represents an option, some of which are scored as having zero effectiveness because of having failed one or more requirements.

RAND TR754-S.3

Next Steps

Our phase-one work demonstrated basic ideas and methods. The key elements of next-phase work should be (1) define ways to make the mission-level analysis more realistic by collecting valid data (to include variations and other manifestations of uncertainty), (2) address a second mission area, and (3) raise the level of analysis to the enterprise level, at which major decisions are made about the Air Force program (pp. 41–43). The sources of more realistic data could include empirical information from operations in the Balkans, Iraq, and Afghanistan; modeling and simulation; interviews with warfighters and other experts (interviews benefiting from the structure provided by our capability model); and interviews with system engineers.

Elevating the work to the enterprise level will require review of the current Air Force planning system (which includes extensive mission-risk analysis), collecting an in-depth understanding of how the current system does or does not deal effectively with warfighter needs, development of alternative procedures, and adaptation of the portfolio-analysis methods to the more strategic enterprise level. One of the most important and difficult challenges is to develop methods to assist decisionmakers in understanding trade-offs *across* missions and capability areas.

Notes

[1]   Our use of "spanning set" should not be interpreted in the literal sense of that term in mathematics, but the concept is qualitatively similar.

[2]   "JTAC" stands for "joint terminal attack controller."

Acknowledgments

The authors appreciate the assistance in the course of this effort by RAND colleague Jody Jacobs; by Lt Col Christopher Chew and Lt Col James Godwin of the Air Force Office of Warfighting Integration and Chief Information Officer, Cyberspace Transformation and Strategy (USAF SAF/XCT); by Air Force consultant Rich Brennan; and by numerous personnel at Nellis Air Force Base who shared operational experiences and provided a good deal of insight and information. These included Maj Scott Campbell, Maj Mike Curley, Maj William Marshall, Maj Kurt Helphinstine, Maj Joseph Campo, Maj Johnny Vargas, Scott Kindsvater, and Russell Handy. Late in the project, we also benefited from discussions with Russell Frasz from the Air Force office that oversees the capabilities-based planning process—the office of the Director, Operational Planning, Policy, and Strategy, Deputy Chief of Staff, Operations, Plans, and Requirements, HQ USAF (AF/A5X)—and with Col George Bochain and Maj Tyson Andersen of the office of the director of Air and Space Operations, HQ Air Combat Command (ACC/A3AF). Finally, we note our appreciation for the careful reviews of the draft manuscript by RAND colleagues Leland Joe and David Vaughan.

Abbreviations

ACC	Air Combat Command
ASOC	air support operations center
BDA	battle damage assessment
C2	command and control
C4ISR	command, control, communications, computers, intelligence, surveillance, and reconnaissance
CAOC	combined or coalition air and space operations center
CAS	close air support
CASEM	Close Air Support Evaluation Model
CRRA	Capabilities Review and Risk Assessment
DoD	Department of Defense
EBO	effects-based operations
GPS	Global Positioning System
HQ USAF A5XC	Air Force Directorate of Operational Capability Requirements
IED	improvised explosive device
ISR	intelligence, surveillance, and reconnaissance
JFO	joint fires observer
JSF	Joint Strike Fighter
JTAC	joint terminal attack controller
MQ-1	Predator unmanned aerial vehicle
MQ-9	Reaper (originally "Predator B")
OSD	Office of the Secretary of Defense
PAT	Portfolio Analysis Tool

SAF/XC	Air Force Office of Warfighting Integration and Chief Information Officer
SAM	surface-to-air missile
SCAR	strike coordination and reconnaissance
SEAD	suppression of enemy air defense
SME	subject-matter expert
SOF	special operations forces
UAV	unmanned aerial vehicle
USD/AT&L	Under Secretary of Defense for Acquisition, Technology, and Logistics

CHAPTER ONE

Introduction

This project came about as the result of an Air Force request to RAND, a request stemming from its concerns that warfighter needs are often not translated effectively into program decisions—despite efforts and processes intended to do so.[1] We were asked to take an independent cut at doing better. If the first-phase effort in doing so was successful, the intention was that we would then relate our suggestions to the current processes and either describe how our suggestions would fit into them or explain what changes in process would be called for.

Diagnoses differ about current difficulties but point to several problems within the Air Force itself. The first is that the leaders deciding on the final Air Force program may not receive adequate information to make informed decisions. That is, by the time some capability options reach late points in the decision cycle, their summarized justification may not effectively explain the value of what is being requested. This is problematic because a cryptically summarized capability regarded as important in one mission area may not be intuitively important to individuals familiar with other mission areas, especially if the capability pertains to something that seems rather mundane (such as improved network capabilities).

A distinctly different problem is that it can be difficult in any organization to move a worthy program through a centralized decision process if the program's benefits are strongest when viewed from a cross-cutting perspective. Stovepiped organizations have difficulty assessing such benefits properly. Figure 1.1, adapted from an Air Staff briefing, is intended to make precisely that point. Within the Air Force, a proposal to improve "Capability X" may have no single natural channel for advocacy because it relates to several of the Air Force's mission panels. The benefit to any one mission area may be significant but undramatic, whereas the overall benefit across mission areas would be seen as substantial if viewed properly. This type of problem can arise, for example, with networking initiatives. Such problems occur in all of the military services, and the General Accounting Office sharply criticized DoD for not having adopted the integrated portfolio methods used in the business world (U.S. Government Accountability Office, 2007). Although DoD is adopting portfolio methods, how they will develop is not yet clear.

Yet another potential cause of difficulty is that Air Force program choices made at a high level, such as headquarters, often attempt to reflect the items on the priority list of the Air Force Chief of Staff. However, such priority lists cannot reasonably include the myriad of lower-level initiatives regarded as important by warfighters in the field. Instead, they tend to address, e.g., big-ticket procurement items or other high-visibility problems. There is a conflict between following priority lists slavishly and developing a balanced program using portfolio-management methods. It is certainly possible and essential to represent priorities in a portfolio-management approach, but it is not so simple as moving down a prioritized list until the money runs out.

Figure 1.1
Cross-Cutting Capabilities Have No Natural Home in Organizational Structure

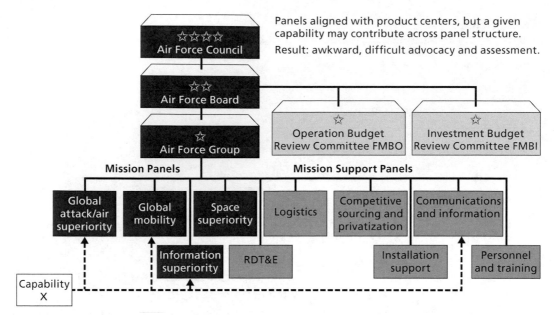

SOURCE: Adapted from Air Staff briefing, September 2007.
NOTES: FMBO = Deputy Assistant Secretary for Budget, Directorate of Budget Operations; FMBI = Deputy Assistant Secretary for Budget, Directorate of Budget Investment; RDT&E = research, development, testing, and evaluation.
RAND TR754-1.1

These concerns about representing warfighter needs led the Air Force to ask RAND's Project AIR FORCE to develop methods for doing better, methods that would use measures of combat effectiveness in Air Force decisionmaking about programs. The approach would also need to include cost-effectiveness work appropriate for enterprise-level processes. It was expected that such work would draw on RAND's considerable work in capabilities-based planning and portfolio-management methods, some of it instigated at the request of the Under Secretary of Defense for Acquisition, Technology, and Logistics (USD/AT&L).[2]

It was agreed that RAND would begin its work with a concrete mission-level example, the mission of close air support (CAS). Although the work would be of a prototype nature, using notional data, the research would benefit from being tied to real-world operational issues. It would also draw on a series of past RAND studies for the Air Force.[3] The disadvantage was in deferring study of enterprise-level issues until the next phase of work.

Over the course of six months, RAND studied the CAS mission, consulted with operators in the field, developed a corresponding analytical model, adapted portfolio-analysis tools, and demonstrated analytic results. The remainder of this technical report documents that work. We begin by describing the generic approach. Subsequent chapters apply the approach to the CAS mission.

A Generic Approach to Mission-Level Capabilities Analysis

Figure 1.2 summarizes the generic approach that we have developed for comparing capability options within a given mission area.[4] The first step is defining the mission area carefully. This might seem straightforward, but is frequently treacherous because mission areas overlap with imperfectly understood seams with other mission areas, and because missions can change over time—often well before official doctrine changes.

The second step (right side of figure) is crucial to analysis under uncertainty. Because the mission may need to be performed in any of many operational contexts, which will affect mission objectives and operational concepts, evaluation of options should be for a "possibility space" or "scenario space," rather than for only a single allegedly standard case. After contemplating the possibility space, it is usually possible to simplify by focusing on a small set of test cases chosen so as to stress capabilities for the mission in all of the relevant ways. An appropriate set of such test cases is called a *spanning set*.[5] Developing an analytically sound spanning set is a matter of both art and science.

Step Three is to understand, define, and model the combat operation. Addressing the various cases may be possible by parameterizing a single model. We refer to such a model as a *capability model*, to indicate that its purpose is to elucidate approximate capabilities for higher-level work in design, resource allocation, and strategic planning—as distinct from, say, simulating an operation in detail, as is often necessary for purposes of training or mission rehearsal, or from modeling in exquisite detail, as is necessary in manufacturing.

Figure 1.2
Approach to Mission-Level Analysis

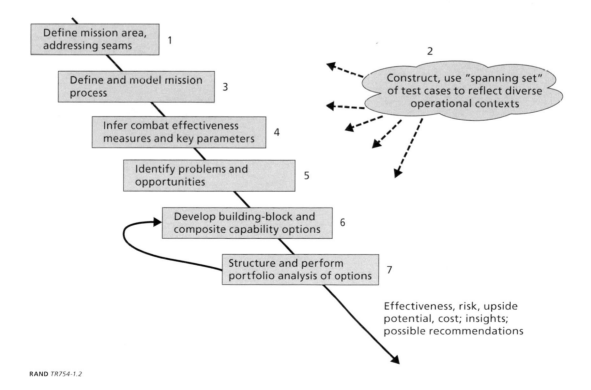

1 Define mission area, addressing seams

2 Construct, use "spanning set" of test cases to reflect diverse operational contexts

3 Define and model mission process

4 Infer combat effectiveness measures and key parameters

5 Identify problems and opportunities

6 Develop building-block and composite capability options

7 Structure and perform portfolio analysis of options

Effectiveness, risk, upside potential, cost; insights; possible recommendations

A well-conceived capability model allows us, in Step Four, to see or infer appropriate combat-effectiveness measures and the parameters on which those measures depend. *The measures suggested are not ad hoc accountant-style metrics, but rather metrics directly related to the ability to perform the actual mission.* This involves understanding clearly the effects to be achieved—both ultimately and along the way. Measures developed in this way tend to be coherent, well behaved, and resistant to misuse or organizational gaming.

Step Five is to identify the operational problems to be solved (i.e., find shortfalls or failure points) and potential improvement options.

At this point (Step Six), it is time to develop options for consideration. Some will already exist; some will need to be constructed. Many of the options that arise have a dependent nature; that is, they may have potential value in an abstract sense, but their operational value depends on other mission-system characteristics. For example, improving communications bandwidth may be useless unless anti-jam capability is also improved. Thus, we combine options treated as building-blocks to create *composite options* that are the appropriate entities to be compared, contrasted, and pitted against each other in the battle for resources. The result is that the options being compared are not posed as some platform versus some weapon versus some intelligence, surveillance, and reconnaissance (ISR) capability, but as different composite ways to accomplish the mission.[6]

The last step (Step Seven) is to construct a suitable analytic structure for portfolio analysis, and then to perform the analysis of the various composite options. The structure used is driven largely by the conclusions of Steps Two and Three about measures of combat effectiveness and test cases. This last step is applied iteratively because of learning and feedback: With the insights from initial analysis, it should be possible to identify new and better composite options.

The remainder of this report describes and applies the methodology. Chapter Two discusses defining the mission area, addressing seams, and constructing test cases. Chapter Three describes developing a capability model and inferring natural measures of effectiveness from the model. Chapter Four discusses the identification of problems, opportunities, and building-block and composite options. Chapter Five walks through the structuring of portfolio analysis and the performance of such analysis on composite options. Finally, Chapter Six draws conclusions and gives recommendations for future work, particularly on (1) operationalizing the application for the CAS mission (e.g., with realistic data) and (2) how to extend the work to the enterprise level. Another aspect of next-phase efforts should be relating our research to the Air Force's Capabilities Review and Risk Assessment (CRRA) process and the considerable data that has been collected within it.

Notes

[1]   We do not discuss the current processes in this report. They include the Joint Capabilities Integration Development System (Joint Chiefs of Staff Instruction, 2009), the Air Force's CRRA process (see Snyder, Mills, Resnick, and Fulton, 2009; Jones and Herslow, 2005), and the new Capabilities Portfolio Management (Department of Defense, 2008). As discussed in Chapter Six, the next phase of effort envisions comparing the current processes with the suggestions made in this report, so as to better assess what improvements might be achieved.

[2]   Much of this has been published in the open domain (Davis, Shaver, and Beck, 2008; Davis, Shaver, Gvineria, and Beck, 2008). See also related work by the Institute for Defense Analyses (Porter, Berteau, Mandelbaum, Diehl, and Christle, 2006; Porter, Bracken, Mandelbaum, and Kneece, 2008).

³ See Pirnie, Vick, Grissom, Mueller, and Orletsky, 2005; Hura, McLeod, Mesic, Sauer, Jacobs, Norton, and Hamilton, 2000; Vick, Moore, Pirnie, and Stillion, 2001; Jacobs, McLeod, and Larson, 2007; Jacobs, Johnson, Comanor, Jamison, Joe, and Vaughan, 2009; and Jacobs, Joe, Vaughan, Dunham-Scott, Jamison, and Webber, 2008.

⁴ The approach adapts past work for the Director of the Missile Defense Agency (Davis, Bonomo, Willis, and Dreyer, 2005; Willis, Bonomo, Davis, and Hillestad, 2006) and prototype analysis of global-strike options for the USD/AT&L (Davis, Shaver, and Beck, 2008; Davis, Shaver, Gvineria, and Beck, 2008). Important elements of the approach were also used in a congressionally requested National Academies study (National Research Council, 2008).

⁵ This should not be confused with the set of official Defense Planning Guidance scenarios of the Office of the Secretary of Defense's (OSD's) Analytic Agenda because those, while dealing with very different political-military cases, do not necessarily stress capabilities systematically in all of the right dimensions.

⁶ Success depends on constructing smart composite options and operational concepts for their evaluation. Doing so may require cooperation among technologists, system engineers, warfighters, analysts, and cost estimators. Some mathematical and computational techniques can help (Davis, Shaver, Gvineria, and Beck, 2008).

CHAPTER TWO

Defining the Mission Area and Challenge Cases

Defining the Mission Area

Let us now apply the approach of Figure 1.2 to the CAS mission area. What is CAS? Defining CAS turned out to be an interesting challenge because its nature has been changing, even though that was not well represented in formal documents as of the time of our study.[1]

According to traditional doctrine (Joint Chiefs of Staff, 2003), CAS is defined as

> air action by fixed- and rotary-wing aircraft against hostile targets that are in close proximity to friendly forces and which require detailed integration of each air mission with the fire and movement of those forces.

Another way to think about the matter is to argue that "CAS is what CAS assets do." But what are CAS assets? The traditional Air Force platform is the A-10 Warthog, which was developed for and is dedicated to the CAS mission and what was once called battlefield interdiction. However, in current operations, the Air Force also employs F-16s and F-15s, as well as unmanned MQ-1 Predators and MQ-9 Reapers. Gunships also perform what amounts to close support. In our prototype study, we focused on core Air Force CAS assets and did not consider use of heavy bombers, Navy and Marine aircraft, or Army helicopters.

Even if we focus on Air Force assets, however, it seems clear that the doctrinal definition needs to be updated because it does not cover important new mission variants that are sometimes called strike coordination and reconnaissance (SCAR), armed overwatch, and corps shaping. Corps shaping was used in the combat phase of Operation Iraqi Freedom; SCAR and armed overwatch are common in today's activities in Iraq and Afghanistan.[2]

As a related matter, thinking about CAS has traditionally focused on combat operations, but planning must distinguish among different phases and classes of conflict: war, stabilization, and counterinsurgency in particular. Thus, the mission area is a good deal broader than might have been thought. If planning is to be forward-looking, representing that breadth is essential.

Finding a Spanning Set of Test Cases

The Range of Possibilities

Table 2.1 depicts roughly the distinctions that arise in itemizing the range of operational circumstances for CAS. This semi-structured table should be viewed as analogous to what analysts might write on a whiteboard while brainstorming the issue. The left column lists the class of factor; the second and third columns itemize important cases or distinctions relating to friendly (Blue) and adversary (Red) forces. Let us go through the items of the table briefly.

1. The first row, on type of operation, reminds us that Blue's CAS operations may be quite different if they are part of defending against a large-scale invasion, conducting limited strikes (perhaps supporting special operations forces [SOF]), conducting offensive maneuvers, or conducting stabilization operations. Red's operations are also different for these cases and may include the CAS-stressful tactic of simultaneous actions at multiple points in the area of operations.

2. The row on political-military considerations is cryptic, but important. For example, worries about collateral damage may be especially important in a particular mission context. Insurgent forces often operate in dense urban areas among the civilian population. They may also move along paths that increase the potential for Blue friendly-fire casualties.

3. Command and control (C2) is always a crucial variable of analysis. Operations are quite different if performed only by the Air Force, jointly with the Navy and Marines, on a combined basis with sophisticated NATO members, or on a combined basis that includes less sophisticated and perhaps ad hoc allies. It is also important to distinguish among different C2 situations. One set of such distinctions is described in NATO's draft concept of a C2 Maturity Model, which refers to levels of maturity as conflicted, deconflicted, coordinated, collaborative, or "edge."[3]

4. Base locations and procedure affect mission performance. The size of the area for which aircraft are responsible, and the base from which they serve that area, affect response times and the density of available assets. Effects will also vary if CAS aircraft operate from steady-state "stacks," are retasked from other duties (such as ISR or strikes of lower priority), or respond from strip alert.

5. The type of "CAS" mission has already been mentioned.

6. Blue's operational objectives should always be to achieve the commander's intended effects, but these effects may or may not be about simple killing of targets. The desired effect may be, for example, to suppress adversary actions, to deter them, or to otherwise influence them (all of these are important in corps-shaping). Further, as mentioned above, a crucial objective may be negative: *avoiding* bad effects. The mirror to this is that Red's objectives may be partly political in nature—encouraging dramatic Blue blunders, such as the killing of civilians that will be perceived as indiscriminate.[4] Yet another operational objective is efficiency, which can pay off, e.g., by allowing broader operations, freeing resources for other missions, and by reducing life-cycle costs.

7. Force sizes, of course, may be large, small, or something in between.

8. Blue capabilities for everything from ISR to weapon delivery are discussed throughout this report and not discussed in this section, but adversary capabilities are important elements of the operating environment, particularly with respect to defenses.

9. Finally (although more attributes of the problem space could certainly be added) are environmental factors, such as weather and the terrain in which the enemy must be located and attacked.

Table 2.1
Brainstorming the Problem Space

Class of Factor	Blue	Red
1. Type of operation	Defending an ally or critical asset from invasion or insurgency Conducting strikes (e.g., with SOF) requiring CAS Conducting offensive maneuvers into enemy territory Conducting stabilization operations	Invasion with large-scale maneuver units Infiltration Insurgency operations such as hit-and-run attacks, use of IEDs Single-point or simultaneous multiple-point action
2. Political-military considerations	Importance of minimizing collateral damage	Desirability of Blue collateral damage
3. Jointness and C2	Air Force alone; with Navy and Marines; with NATO allies; etc. Conflicted, deconflicted, coordinated, with collaboration, on the "edge"[a]	Disorganized, organized, organized with cooperation, etc.
4. Bases, procedures	Proximate to operating area versus distant Classic "stacks" versus strip alert, etc.	Basing in open areas, forests, mountains, dense urban; isolated or immersed in civilian structure
5. Type of CAS	Classic close support Close support of SOF Corps-shaping Armed overwatch	N/A
6. Operational objectives	Effects via timely kill, suppression, or influence of enemy forces or targets Avoiding negative effects	Classic military effects versus, e.g., achieving drama by forcing errors and indiscriminate killing by Blue Efficiency
7. Force sizes	Large, small	Large, small
8. Capabilities	As discussed elsewhere in report	Air defenses Electronic countermeasures Passive defenses
9. Environment: weather, terrain	Weather, terrain (affected by Red's tactics [see 4])	Weather, terrain

NOTES: IED = improvised explosive device; N/A = not applicable.

[a] This breakdown borrows from the draft C2 Maturity Model, which depicts the maturity of the C2 system with five levels (NATO Command and Control Research Program [SAS-065], 2009). The "edge" assumes a high degree of delegation, as well as local initiative, trust, familiarity, and self-synchronization.

An Approximate Spanning Set of Cases

Table 2.1 conveys a sense of just how diverse the CAS mission may be. To be comprehensible, however, analysis needs to limit dimensionality. Upon thinking about the dimensions in Table 2.2, we concluded that the primary stressors of CAS, i.e., the ones on which we should focus analytic attention (aside from Blue capabilities), were (1) air defenses, (2) countermeasures against communications and weapons, (3) target density and number and the area of operations, (4) detectability and vulnerability of targets, (5) collateral-damage and fratricide-related constraints, and (6) timeliness required. Other factors are also important, but were not needed for our prototype work.

One point does need elaboration, however. This discussion of stressors is cast in the usual language of air forces, i.e., in terms of killing targets. Regrettably, this does not convey a sense of the connections with ground warfare, as in protecting friendly forces or countering enemy maneuver. These crucial relationships are discussed more in Chapter Three.

Table 2.2
A Spanning Set of Scenarios for CAS

	CAS Mission Type and Scenario	
Difficulty Version	Maneuver: Near Peer or Rogue	Stabilization: Post-Combat Insurgency
Low	Airfields fairly distant from targets; sizable numbers of targets; some jamming and SAMs	Fewer density of targets, but those may be close to or comingled with civilians
High	More significant jamming, more SAMs; higher target densities	More difficult target identification; more serious communications and weapon-system interference, more serious SAM threat

NOTE: Cell entries highlight issues addressed in the "scenarios" used to depict the related case.

Although a considerable simplification, Table 2.2 suggests that most of the issues of interest for investment planning could be encapsulated in four cases. These distinguish between variants of the CAS mission that focus on supporting large maneuver units and those that focus on stabilization-phase operations combating insurgency (columns). We also distinguished between low- and high-difficulty cases (rows). Although the cases do not mention specific threats, we considered specific possibilities when interpreting the abstract cases. Adversaries, for example, might be North Korea, Iran, insurgents in Iraq or Afghanistan, and so on. In distinguishing between low- and high-difficulty cases, we were anticipating the potential consequences if, for example, our adversary had large numbers of relatively effective man-portable surface-to-air missiles (SAMs) or the capacity to perform communications or jamming of the Global Positioning System (GPS).

Specialization for the Prototype Problem

Figure 2.1 indicates schematically how we defined elements of the system. In a classic depiction of CAS, both sides have maneuver units, but we can include cases where the adversary "units" are small groups of irregulars conducting hit-and-run operations on friendly forces or, e.g., significant installations or cities. We assumed that CAS aircraft maintain a "stack" from which aircraft are dispatched as necessary to deal with enemy targets that arise. Joint terminal attack controllers (JTACs) are an essential element; they direct fires and minimize risk to friendly forces and civilians. (JTACs could be supplemented by other forward observers.) C2 includes sortie-generation planning and priority-setting at the combined or coalition air and space operations center (CAOC) level and at the air support operations center (ASOC) level, assigning aircraft to targets and granting or denying permission to attack targets. Attack permissions may be delegated farther down the line to at-scene forward observers. This terminology applies to both combat and stability phases of operations in theaters such as Iraq.

Figure 2.1
Illustrative "Traditional" CAS Mission

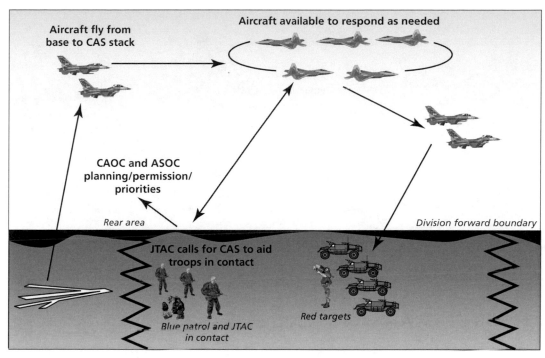

RAND *TR754-2.1*

Notes

[1]   See various official documents (U.S. Air Force, 2003, 2007; Joint Chiefs of Staff, 2003). Some publications describe more recent operations and draw contrasts with the past (Kirkpatrick, 2004). A number of RAND studies bear on CAS issues (Pirnie, Vick, Grissom, Mueller, and Orletsky, 2005; Hura, McLeod, Mesic, Sauer, Jacobs, Norton, and Hamilton, 2000; Vick, Moore, Pirnie, and Stillion, 2001; Jacobs, McLeod, and Larson, 2007). A recent study for the Air Force Chief of Staff drew strong conclusions about Air Force initiatives needed to improve effectiveness in irregular warfare (Mesic, Thaler, Ochmanek, and Goodson, 2010).

[2]   Discussion can be found, for example, in several recent publications (Grant, 2008; Kirkpatrick, 2004; Third Infantry Division [Mechanized], 2009).

[3]   This is based on a draft report currently being peer-reviewed within the NATO community of experts (NATO Command and Control Research Program [SAS-065], 2009). Final publication is anticipated in early 2010.

[4]   Alternatively, one can regard the need to avoid bad effects as a constraint.

CHAPTER THREE

A Capability Model for CAS Mission-Area Analysis

Prefacing Comments

Capability Models in the Larger Hierarchy

The capability models that we recommend can be understood by viewing their relationship to other models, as in Figure 3.1. The figure shows the classic levels that have long been used to distinguish among military models (particularly for Air Force applications): campaign, mission, engagement, and engineering. Models at each of these levels can be relatively simple or complex. A high-resolution mission-level model may follow the individual aircraft of a day's air operations as they fly to their target areas, avoid or interact with ground-based air defenses and enemy aircraft, drop ordnance, and return. The level of detail may include following, moment by moment, whether the various aircraft are being illuminated by defender sensors, the altitudes at which they are flying, their spacing, and so on. A simplified mission-level model, how-

Figure 3.1
Capability Models in the Hierarchy of Model Types

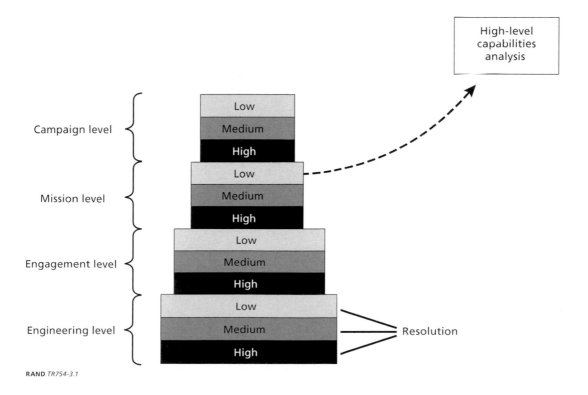

ever, may follow a representative aircraft and merely distinguish coarsely among periods such as ingress, weapon delivery, and egress—characterizing what happens in those periods with concepts such as probability of survival.

Ultimately, analysis should draw on diverse models, human games, and other sources of information, such as history. That is, good analysis should draw on a suite of tools. Low-resolution work should be informed by lessons from higher-resolution work, and vice versa. This is not a mere admonition for the record, but rather a matter of some importance.[1] Nonetheless, in what follows, we focus on relatively low-resolution models.

Capability models—the models that are most useful in higher-level capabilities analysis and in higher-level systems engineering for early design and exploratory trade-offs—are often at the mission level but have low resolution and are relatively simple. They are highly parameterized, which allows users to explore a wide range of cases easily. They are transparent, so that users can understand the results. These are the models of most use in the type of strategic planning we have in mind (i.e., headquarters-level force planning about future capabilities and force structure). Such higher-level work is concerned about uncertainties and the corresponding management of risk. What is needed is not a single high-resolution simulation based on an approved set of data (input assumptions), but an understanding based on exploratory analysis over the range of uncertainties (and disagreements). For that, the simpler capability models are quite powerful. As noted above, however, using them well may require analysts to understand phenomena well at higher levels of resolution, for which they may use more detailed models as well. That is, overall, analysis may employ multiresolution modeling.[2]

Purposes of the Capability Model

Understanding and Representing. For capabilities analysis of the CAS mission area, then, we sought a relatively low-resolution, analyst-friendly model. The first and foremost reason was that such a tool can help us "write down" what we learn so as to understand the operation itself. Modeling forces us to have definitions, distinctions, and an understanding of relationships. A good simple model with such attributes improves communications, allowing us to reason together—not only among modelers and analysts, but with warfighters on the user end and officials at the resource-allocation end.

A capability model should represent the phases of real-world operations and, consistent with taking a system view, should highlight the *critical components* of the operation. Often, a system fails unless a number of its components all work properly. That is, each such component is critical, not just "nice to have." Improving an already-good radar cannot compensate for a weapon that fails to detonate; improving the lethality of an already good weapon cannot compensate for an inability to find a target in the first place, or for an inability of the platform to penetrate air defenses so as to be able to engage a target. Clearly, investments on the margin should reflect this system thinking. Superficially attractive options may do literally no good unless investments are also made in other things, even if these other things are apparently mundane and not explicitly on the priority lists of policymakers. This reality is well known to warfighters, but it is commonly dealt with poorly in the development and execution of programs. For example, in the 1980s and into the 1990s, the military services consistently invested far too little in precision-guided munitions and intelligence-support packages, thereby undercutting the effectiveness of air-delivered strikes.[3,4]

Natural Measures of Combat Effectiveness; Metrics. If the modeling is done well, with an understanding of the operations, good measures of combat effectiveness fall out naturally.

Instead of metrics coming about as someone's ad hoc "bright ideas," the measures relate directly to the objectives of operations and the effects they are intended to achieve (Kelley, Davis, Bennett, Harris, Hundley, Larson, Mesic, and Miller, 2003). This is no minor issue, since people often use counterproductive metrics that are not well connected to operations or that give misleading impressions.[5] Suppose that a commander seeks to maneuver friendly forces to a strategically important location in a theater. He may wish to do so as rapidly as possible to avoid preemptive enemy action and to preserve his resources for subsequent activities. Consider two cases of how matters might play out. The first case is unfortunate because en route battle is not avoided, delays occur, and casualties are taken. The Blue force wins the battle and continues to its destination, but not as intended and potentially too late. In the second case, the maneuver is fast and without the diversion and price of en route battles. Clearly, the second case is good; the first is bad. Now, however, suppose that we were evaluating the effectiveness of supportive CAS operations. A naive set of metrics for CAS would include, prominently, kills per day by CAS aircraft. By this metric, the first case would be good and the second case would be bad!

Capability Experiments. Given a model and good effectiveness measures, we can conduct "capability experiments" to understand the benefits that would accrue with improvement options. Even more important, we can understand *why* the benefits occur (or do not materialize as expected). Part of the benefit of the experiments is to find which capabilities, as proxied by which model parameters, really matter.

A number of other considerations apply. First, the model must allow us to represent improvement options in its parameters or it is not useful. No model, however, can cover all the issues, so side-study issues need to be identified.

Explanation. A final function of modeling is to understand the problem so well as to permit using an even more simplified model in the final portfolio analysis—one that should be readily understandable and yet tell a correct and compelling "story."

Designing a Capability Model for CAS: Relating Target-Killing Capabilities to Ground-Commander Needs

Requirements

Against that background of technical philosophy, let us next ask what the essence of a capability model for CAS should be. We concluded that a single principle should be front and center:

> First, all good measures of effectiveness are focused on *output*, which is best judged by the customer—in this case, the Joint Commander and subordinate Army commanders.

The issue is not whether the Air Force believes a level of CAS capability is good and efficient based on convenient measures, such as sorties flown; the primary issue is, How is it viewed by the Army commanders who depend on air forces?[6]

Translated, this means that CAS support must not only be effective on average, but must be *timely and reliable.* To a ground commander, what matters is achieving the appropriate effects, which may be suppressing or almost always responding to adversary attacks of exposed Army units *before* they suffer unacceptable attrition. Or perhaps a corps commander wants shaping operations that will assure that key bridges are temporarily closed to prevent attacks on his maneuver units at a specific time tomorrow, but with the bridges repairable for friendly use.

Such effects are not captured well in such aggregate statistics as sorties flown or even enemy vehicles killed per day. Equally important is avoiding "bad effects," which can have strategic significance. These include fratricide (including casualties to allied units), collateral damage (e.g., bombing a wedding party instead of an insurgent safehouse), and minimizing attrition of CAS assets.[7]

Though the focus should be on desired effects, efficiency also matters, as mentioned earlier. Efficiency permits doing more with the same assets and frees funds for other purposes. In this study, we treated efficiency as one component of effectiveness for the sake of compactness.

We raise this matter explicitly because it poses problems for analysis. How does one measure ability to deter or suppress? We concluded that the *capability* to kill a sizable fraction of targets in a timely manner, as measured by a scenario with high target density and target diversity, would also imply the ability to deter or suppress the maneuvers of an intelligent adversary. That is, success might mean *no* kills (because the adversary was deterred from conducting operations). A bit of reflection will convince the reader that measuring combat effectiveness empirically requires a good deal of thought rather than mere number-crunching of readily available data.

Although we do not pursue the issue in this report, Figure 3.2 suggests how the effects achieved may depend on the level of CAS capability (measured by the damage to the enemy on the x-axis). The concept here is that if the adversary knows he will suffer higher and higher levels of damage, he is more likely to be deterred. However, how much damage he can accept depends on how important the maneuver is to his operations (hence the two curves). The concept suggested in the figure is an important connection between analytical work to assess CAS effectiveness, as in the remainder of this report, and achieving the actual effects desired.

Figure 3.2
Deterring Maneuver with CAS Capability

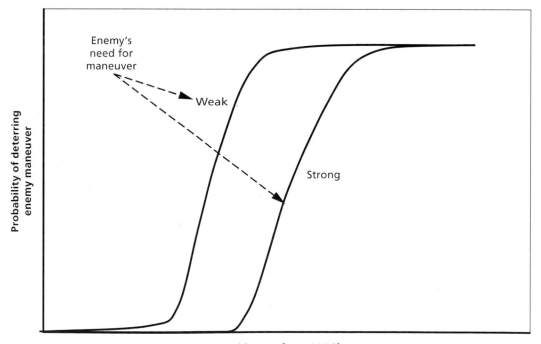

The same kind of reasoning can be applied when attempting to measure air forces' *capability* to protect friendly ground forces and shape the battlefield for the ground commander (e.g., by deterring, deflecting, or destroying enemy maneuver units that would otherwise interfere with friendly operations). Again, the concept here is that the capability (i.e., the wherewithal and skill) can be measured by estimating the ability to attack a diversity of targets *in a timely and effective manner*. If that capability is good enough, then air forces will be able to conduct the operations of concern to ground commanders.[8]

A Family of Capability Models

We developed a small multiresolution family of models for our work. The simplest of these calculates effectiveness as a product of probabilities. A somewhat more detailed version decomposes those probabilities further but is still a "formula model." The most detailed model (the Close Air Support Evaluation Model [CASEM]) is a stochastic simulation. The stochastic model is important to understanding the dynamics of the mission and because overall effectiveness is ultimately a complicated function that is not necessarily well approximated by a simple product of probabilities, at least not without some adjustments.[9] Figure 3.3 describes the most important features of our simplified capability model schematically.[10]

The lower portion of Figure 3.3 relates directly to the stochastic simulation CASEM, the core logic of which is described schematically in Figure 3.4 and in Appendix A. CASEM is stochastic to allow for the random manner in which targets appear on the battlefield, remain vulnerable, and have time-sensitivity (i.e., must be attacked quickly to serve ground commanders' needs); in which shooters must be reallocated as time-sensitive targets arise; and in which the air operation generates and maintains a stack of aircraft in the general target area. The

Figure 3.3
Structure of the Capability Model

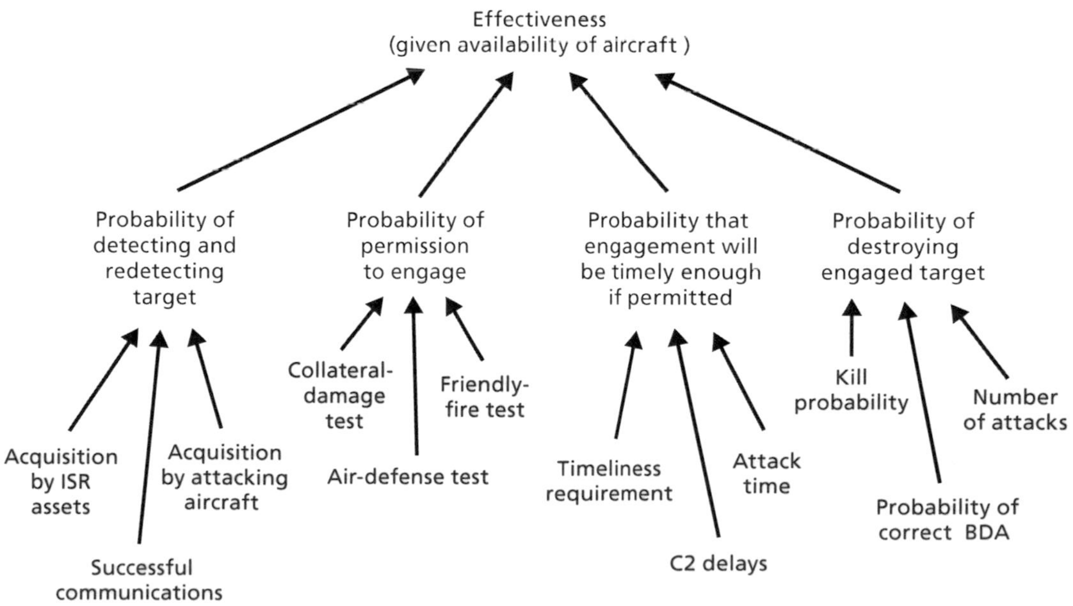

NOTE: BDA = battle damage assessment.
RAND TR754-3.3

Figure 3.4
Core Logic of the CASEM Simulation Model

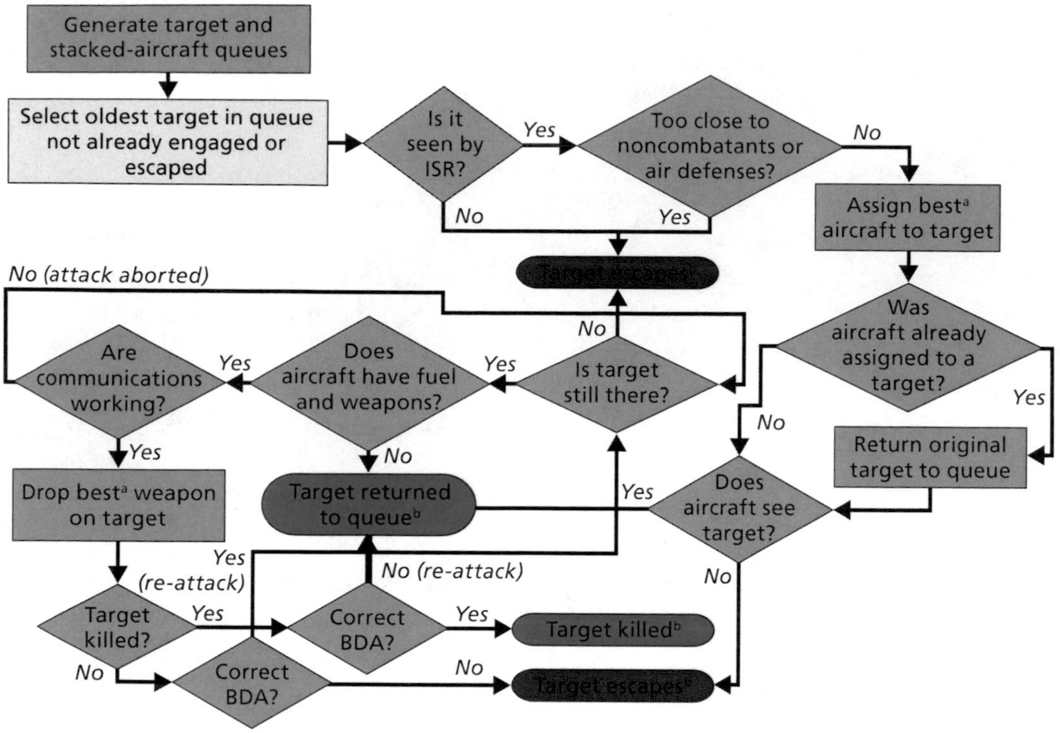

[a] Best = maximizing expected value of targets destroyed.
[b] Model cycles through targets until all targets are resolved.

RAND *TR754-3.4*

CASEM model itself is implemented in a Microsoft Excel spreadsheet using Visual Basic for Applications.

Inferring Natural Measures of Effectiveness

An important element of our methodology is inferring measures of effectiveness from an appropriate capability model, rather than creating them ad hoc. Natural measures can be seen by reading across Figure 3.3 at the second level. We see that CAS effectiveness in the classic mission is said to be dependent on (1) the probability of a target being observed by "the system," whether by ISR assets or troops in contact, and subsequently by the attacking aircraft (if an attack is ordered); (2) the probability that the shooters will be permitted to engage an observed target (not a certainty because of worries about collateral damage, fratricide, and defenses); (3) the probability that a target that is engaged will be struck soon enough to accomplish the mission satisfactorily (e.g., in time to protect ground-force units or before the target disappears); and, finally, (4) the probability that the shooter will actually destroy the target if permitted to attack and able to do so on a timely basis. An additional factor, not shown in the figure, is the probability that aircraft will be available to make the attack.

In this case, the top-level model is as simple as one could hope for: The overall effectiveness is computed by merely multiplying the factors together. Thus, effectiveness is zero if *any* of the factors are zero; if all of the factors are one, then that circumstance is sufficient for perfect effectiveness. In other cases, effectiveness scales with the individual factors: A degradation of 25 percent in any of them implies a 25 percent diminution of overall effectiveness.[11]

As indicated lower in the chart (red lettering), the next level of detail is still relatively simple. Nonetheless, it has more detail that is important for understanding the operational process. Each of the nodes is the basis of a lower-level measure of effectiveness. Here, the factors do not simply multiply. For example, a delay in C2 decisionmaking is but one of the ways in which the effort to engage a target would be too late to be considered successful. It might be that the target was present only briefly, that decision was delayed,[12] that an aircraft was not available, or that the time to make the physical attack was too long. The leftmost two factors of the second level are merely products of the lower-level factors.

Note that using the measures of effectiveness arising as described above is "natural" in that they have a distinct home in our understanding of the phenomenon as reflected in the model. Further, such measures relate clearly to intended "output" (i.e., mission effectiveness). That said, the value of improving a given measure depends on the status of the other factors. That is, overall effectiveness is a "system issue," requiring that one pay appropriate attention to all the critical factors. Our emphasis on "natural measures" is in contrast to the common practice of metrics being identified in advance and then imposed, sometimes with counterproductive results that include organizations learning to game the metric, overemphasis on only portions of the overall problem, or worse. One reason that the Air Force has come to focus on achieving "effects" is because, in the past, there was a tendency to instead focus on inappropriate measures, such as sorties per day or tons or ordnance delivered per day.

A related matter in thinking about measures involves explanation and reasoning. A primary purpose of a capability model is explanation. Its key variables should help "tell the story." Analysts need to understand *why* results are as claimed—both to assure that the model is doing what was intended and, subsequently, to understand the phenomenon. The measures of effectiveness used should fit naturally into such explanations.

Figure 3.5 illustrates the use of the model to understand the CAS "mission system." Assuring high capability levels means addressing many different potential failure modes; this chart shows how, in a given set of runs for a particular scenario, about 75 percent of the targets escaped. About 14 percent escaped because the target was not detected in the first place; another 14 percent were initially detected and an aircraft assigned, but the aircraft was not able to reacquire the target. In a very few cases (almost zero), there was no available aircraft. Some 45 percent of targets could have been engaged but escaped because final permission to attack was withheld because of concerns about collateral damage, fratricide, proximity to noncombatants, or local air defenses with range greater than aircraft standoff ranges. In this scenario, at least, target escapes due to diversion of aircraft were not a problem[13] and, if attacked, targets were almost always killed and BDA was not a factor. Obviously, under different sets of assumptions, all of these numbers would change. The point is merely to demonstrate the explanatory features of the model. This kind of information (were it based on solid data) would suggest where remedies are needed.

As a second example of how we used the model, Figure 3.6 compares the efficiency achieved by CAS in two cases. It compares results with baseline capabilities and with improved capabilities as the result of a composite investment option addressing multiple issues: improved standoff range, communication reliability, and target acquisition.

Figure 3.5
Illustrative Explanatory Output of Model: Why Targets Are Not Killed (notional data)

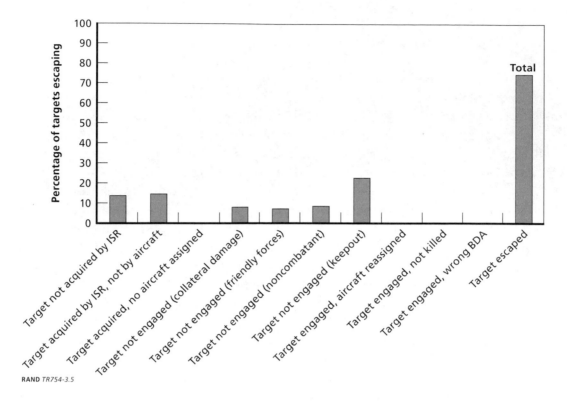

RAND *TR754-3.5*

Figure 3.6
Comparing Efficiency Across Cases

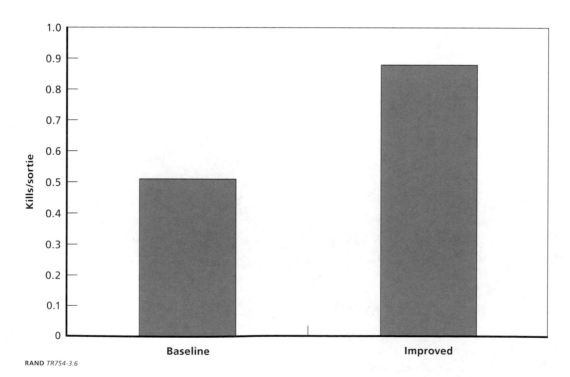

RAND *TR754-3.6*

It is typical in such work for the good options to be composites: It may do little good to fix one problem when there are numerous opportunities for failure. Analysis can help find those good composite options. Results can be surprising; for example, in some of our analysis, we considered doubling the number of CAS aircraft available, only to find that effectiveness was not improved at all. Efficiency, moreover, dropped. That type of result is not uncommon in systems work, where problems may not be solvable by brute force (e.g., problems limited by C2, surveillance, or doctrine).

Notes

[1] Several discussions of the family-of-tools approach are available (National Research Council, 2005; National Research Council, 2006; Davis and Henninger, 2007; Davis, 2002; Hughes, 1989).

[2] See Davis and Hillestad, 1993, and Davis and Bigelow, 1998 and 2003. The centrality to capabilities-based planning has been elaborated (Davis, 2002; Johnson, Libicki, and Treverton, 2003).

[3] See a contemporaneous RAND study (Bowie, Frostic, Lewis, Lund, Ochmanek, and Propper, 1993).

[4] Doctrinal recognition of such system problems is the DOTMLPF process, which emphasizes the need to address doctrine, operations, training, materiel, leadership and education, personnel, and facilities.

[5] Notorious examples include sorties per day and tons of munitions dropped per day. In the 1991 Gulf War, the Air Force and Navy flew very large numbers of sorties, but a disproportionate share of kills was made by a small number of aircraft with precision-weapon capability.

[6] This should be tempered by noting the economist's crucial admonition that "free goods" lead to gross inefficiency. Thus, we have in mind sensible Joint and Army commanders.

[7] The reader will notice relationships to effects-based operations (EBO), which are discussed in many publications (Deptula, 2001; Deptula, 2006; Davis, 2001; Dubik, 2003; Elder, 2006; Grossman, 2006; Smith, 2006; Van Riper, 2006; Davis and Kahan, 2007; Jobbagy, 2006), most recently in a very critical interpretation by the Commander, Joint Forces Command (Mattis, 2008). We do not discuss EBO here except to note that EBO's *better* interpretations relate well to our discussion. Unfortunately, EBO has become associated with confusing use of the English language, circuitous thinking, ponderous planning processes, and dependence on complex and dubious operational net assessments that are inappropriate for rapidly changing operational circumstances.

[8] An alternative approach might be to define rich and detailed scenarios in which air forces must conduct particular counter-maneuver, corps-shaping, or friendly force–protection operations, and to then use simulation outcomes as the measure of capability. Simulating combat in such scenarios would be useful for *illustrations*, and might be quite effective in some studies and exercises. The approach we have taken, however, seeks to provide a simpler measure of the flexible, adaptive capability that would permit successful operations in a wide variety of CAS circumstances. Its appropriateness depends on using target-generation statistics (diversity of target types popping up over time, with appropriately varied requirements for response time).

[9] The significance of the statistical effects is illustrated with good examples in an article by Thomas Lucas (2000). With an understanding of the underlying statistics, however, it is often possible to use the simpler calculations by evaluating the component probabilities using, e.g., means or medians rather than most-probable values.

[10] Analysts will probably regard this model as logical and uncontroversial. Note, however, that the models often used for capabilities analysis are complex computer simulations in which these driving factors are not readily visible. Indeed, most of them are *outputs* of the simulation. We believe that our simpler approach is superior for higher-level capabilities analysis, whereas the more complex simulations are far more powerful for other purposes, such as integrative work. See also National Research Council (2005), which discusses good practices.

[11] In technical terms, the higher-level model is a *motivated metamodel* (Davis and Bigelow, 2003). Its structure was based on physical intuition about the problem, but it has been calibrated via statistical analysis to results of the CASEM simulation. See Appendix B.

[12] The need to minimize delays and ways to accomplish that are discussed in Jacobs, McLeod, and Larson, 2007.

[13] In most of our work on CASEM, we scaled the sortie generation rate with the target presentation rate so that we ended up with roughly the right number of aircraft. The assumption here was that the commander would know, roughly, what was needed. For the particular analysis, we did not want the number of available sorties to be the main factor affecting the results. Had we been simulating an entire battlefront and force structure, different assumptions would have been necessary.

CHAPTER FOUR

Illustrative Capability Options

Given an understanding of the mission and a model with which to represent it, the next step is to develop and characterize some options to improve warfighter effectiveness. We focused strictly on improvements of Air Force systems and procedures. The examples we identified come from capability experiments with the model, field research to discuss and observe issues with CAS pilots and controllers at Nellis Air Force Base, study of documents provided by the Air Force from planning exercises, and general research.

Identifying Problems

Table 4.1's left column summarizes an illustrative set of problems that we uncovered to which CAS-related options are relevant.[1] Many other even more mundane but important equipment problems exist. Radios, for example, are unreliable; too many radios have to be carried because of interoperability problems; and the radios are heavy. The airmen on the ground who must carry this equipment are seriously overburdened.

Many problems relate to operations. C2 issues are sometimes serious, undercutting strong joint support of the warfighter. Some Air Force operators that we interviewed observed, with chagrin, that in some respects the Marines handle the CAS mission better than the Air Force. Air Force processes and doctrine are sometimes problematic. Fortunately, airmen in the field go around doctrine to get the job done whenever possible, but doctrine needs to catch up. This includes addressing the problem that the Air Force does not have as many JTACs/joint fires observers (JFOs) as appear to be needed for the ground fight, with the Army's increasing dependence on small-unit operations.[2] Collateral damage and fratricide are constant worries because of their strategic significance.

Finally, there are problems of threat. For example, jamming of radios or even of the GPS can occur, and the potential exists for large numbers of shoulder-fired SAMs at unknown locations and with increasing range. CAS is probably not even feasible until after suppression of enemy air defense (SEAD) operations have effectively degraded more-capable SAM systems, eliminating even the threat of a wily enemy who minimizes broadcasts.

Other problems exist, but these are illustrative. They have been chosen in part to show examples of how solutions can take very different forms.

Table 4.2 provides cryptic descriptions of the options mentioned in Table 4.1. These are all of the "building-block" variety. That is, they are low-level options that in many cases make sense only in combinations, because fixing one problem may do no good without fixing another problem at the same time.

Table 4.1
Problems and Building-Block Options to Mitigate Them

Problem	Building Block(s)
Communications Equipment	
Communications jamming, due either to Red efforts or to interference with Blue systems	Improved anti-jamming capability
JTAC and aircraft must quickly "get on same page" about Blue, Red, and collateral damage concerns	JTAC has own-position digital marking; shared sensor point of interest; "John Madden" telestration; full-motion video
Buildings and terrain block call for support, JTAC-CAS aircraft communication	Beyond line-of-sight voice communications
Force Structure and Organization	
Not enough JTACs for maneuver units in modern Army doctrine	Additional JTACs
JTACs cannot be present at point of each attack	Additional JFOs
Targets surprise Blue and require CAS under undesirable circumstances	Additional unmanned aerial vehicles (UAVs) on overwatch
Weapon Systems	
Weapons difficult to tailor for fratricide and collateral damage risks	Dial-a-yield munitions
Air defense systems threaten CAS aircraft	Standoff precision munitions
Uncertainty about Blue force location increases fratricide and time spent managing risk of fratricide	Total Blue Force Tracking
C2 Doctrine	
CAOC does not respond promptly to air support requests routed through it	Increase ASOC autonomy
Acquisition Procedures	
CAS community has to go through lengthy process to fund cheap, basic, common-sense improvements that have little appeal at headquarters when compared with, e.g., F-22s, Joint Strike Fighters (JSFs)	Discretionary funds for the CAS community

Characterizing the Options for Analysis

The preceding discussion assumed that we could assess the attractiveness of the options with our capability model. That, however, depends on translating the options into parameter values of the model. As suggested by Figures 4.1 and 4.2, there are several ways to think about how to do so. Most of the building-block options can be classified in categories such as ISR, Communications (i.e., communication systems), Platforms, and Weapons. Model parameters related to these Blue capabilities are affected by the building-bock options, as suggested in Figure 4.1. However, another way to view things is to consider what phase of the mission is affected (Figure 4.2) and how the building-block capabilities would affect "scenario parameters" of the model, e.g., parameters such as resistance to enemy countermeasures (e.g., communications jamming).

Table 4.2
Capability Options

Building Block(s)	Description
Improved anti-jamming capability	Prevents attack failure due to loss of voice communications as the result of jamming or frequency interference
JTAC own-position digital marking and shared sensor point of interest; "John Madden" telestration	Pilot sees digital representation of JTAC's position; JTAC sees where pilot's targeting pod is pointing. The speed and accuracy of general-area orientation and target talk-on are improved.
"John Madden" telestration	JTAC can mark and share digital image with pilot, highlighting target, Blue forces, and/or collateral damage concerns. The speed and accuracy of general-area orientation and target talk-on are improved.
Full-motion video	JTAC can see what pilot sees through targeting pod and indicate when pilot has target in crosshairs. The speed and accuracy of general-area orientation and target talk-on are improved.
Beyond line-of-sight voice communications (e.g., via satellite)	JTAC (or other parties at point of engagement) do not have their communications inhibited by terrain or distance, and can reach the aircraft (or ASOC) more rapidly
Additional JTACs	More JTACs increases capacity to handle CAS engagements; performance in average engagement is enhanced as likelihood that JTAC will perform Type 1 CAS increases
Additional JFOs	JFOs enhance performance in average engagement, working effectively with JTACs to facilitate CAS engagements at which the JTAC is not physically present
Dial-a-yield munitions	Aircraft can tailor desired effects, lowering number of CAS requests that cannot be satisfied because of fratricide or collateral damage concerns
Standoff precision munitions	Aircraft can stay outside threat envelope of more air defenses, lowering the number of CAS requests that cannot be satisfied because of air defense concerns
Total Blue Force Tracking	Assured awareness of Blue forces lowers the number of CAS requests that cannot be satisfied because of uncertainty about Blue location, and lowers the likelihood of fratricide
Increase ASOC autonomy	Greater ASOC autonomy eliminates important C2 friction and increases responsiveness
Discretionary funds	CAS community can fund cost-effective improvements to CAS, taking advantage of opportunities that would otherwise not have been funded quickly, if at all[a]
Additional UAVs on overwatch	Greater numbers of UAVs improve situational awareness, raising probability that targets will be identified and passed to supporting strike aircraft and lowering probability of fratricide or collateral damage[b]

[a] The value of discretionary funds was not, of course, assessed with our capability-model approach. However, we have included it as an important placeholder for administrative options that can improve capability.

[b] We did not evaluate armed UAVs explicitly in our prototype work, but in a full analysis they would be a priority item for examination.

Constructing Composite Options

Given a good set of building-block options, which typically are generated from many different sources, it is important in program development and analysis to construct packages of such options that make sense and are described in terms appropriate to higher-level decisionmaking. We call these *composite options*. They may include combinations that no one thought to consider before. Although we did not do such an exercise in the prototype study, the approach that we recommend for constructing composite options is to use a procedure that we have docu-

Figure 4.1
Relating Building-Block Options to Analytical Parameters

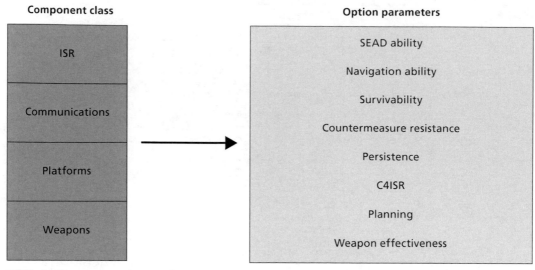

NOTE: C4ISR = command, control, communications, computers, intelligence, surveillance, and reconnaissance.

RAND *TR754-4.1*

Figure 4.2
Relating Options and Parameters

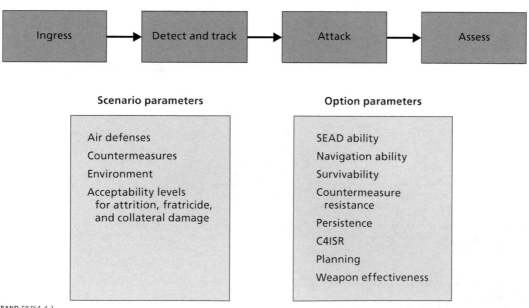

RAND *TR754-4.2*

mented elsewhere.[3] Basically, one programs the computer to construct all possible combinations of the building-block options and then uses a simplified filtering analysis to identify those combinations that are worth considering further.

Unlike more classic methods, the procedure looks for composite options that are reasonably close to the so-called Pareto-Optimal Efficient Frontier in a plot of one-dimensional effectiveness versus cost.[4] Points on the frontier are such that all points inside the frontier are less effective for a given cost or more costly for the same effectiveness. In Figure 4.3, the points represent alternative options. Point A is on the frontier, Point B is close enough to be retained, and Point C is not competitive (i.e., it is "dominated" by Points A and B). That is, it would be discarded without further evaluation. Given uncertainties in estimates of both cost and effectiveness, it may turn out, eventually, that Point B is superior to Point A.

The other important and novel feature of our approach is that we recognize that where the options fall on such plots depends on how one estimates "net effectiveness." Such estimates invariably oversimplify what should be a multicriteria assessment. We refer to a particular way of calculating the net effectiveness as representing a *perspective* because it may reflect judgments or values about the relative effectiveness of various criteria.[5] Uncertainties and disagreements on such matters dominate many of the calculations, so we construct corresponding plots for alternative perspectives. Then we select composite options that rank well (are at least close enough to the efficient frontier) in at least *some* of the relevant perspectives. This is in contrast to seeking only those options that rank well across all of the perspectives. Our reason for this is that we wish to avoid prematurely discarding options that may prove to be competitive in a fuller analysis. After all, initial screening may occur at a time when the criteria levied (i.e., the

Figure 4.3
Finding Good Composite Options

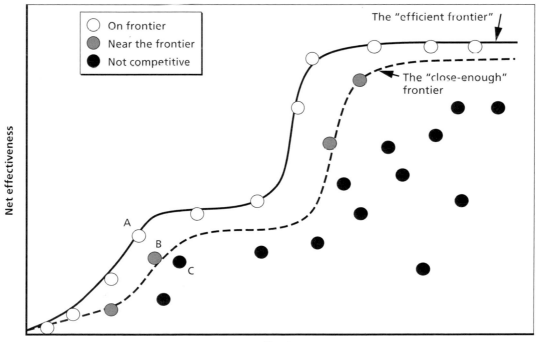

"requirements") exceed what can actually be accomplished or what policymakers will eventually be willing to pay for. We do not want to drop options merely because they don't do well in *all* of the relevant perspectives.[6]

Some of the resulting composite options are likely to be familiar or to be generated quickly by intuition. Others are likely not to have been considered, for reasons that include the natural parochialism of organizations, the past-experience-based blinders that can hamper even very good analysts, and so on.

After the candidate composite options are generated, the analyst should scrutinize the computer-generated candidates and adjust accordingly. Some of the computer's suggestions will be foolish for reasons known to the analyst but hidden from the filtering algorithm; others will be so implausible technically or organizationally (or even politically) as to not be worth pursuing. It may also happen that some of the options that do not appear to be good candidates should be restored to the list because, for one reason or another, they need to be considered. For example, they may have strong virtues ignored by the computer's analysis (or be favored by someone important). This man-machine interaction should be considered a strength, not a limitation.

Because of limited time and resources, we did not literally use the above procedure in our prototype effort. Rather, we drew on intuition and, significantly, on documentation from Air Force exercises and discussions with Air Staff officers to identify some plausible candidates for composite options—candidates adequate to illustrate portfolio analysis in the next chapter. Table 4.3 summarizes those composite options.

Notes

[1] That is, we do not include options for, e.g., improved SEAD. Those are relevant to CAS, but not to CAS programs.

[2] See Jacobs, McLeod, and Larson, 2007.

[3] Theory and an implementing program (BCOT) are described elsewhere (Davis, Shaver, Gvineria, and Beck, 2008). BCOT is still "beta software." The next step in development should include tightening and testing it so that it can be made more generally available. In a parallel effort, one of us (Dreyer) developed a prototype genetic algorithm program for the same objective as BCOT. It worked well and would have advantages for large numbers of building-block options and criteria for evaluation.

[4] Pareto optimality is discussed in any of many basic texts in operations research or economics, or, e.g., in Wikipedia. See also Davis, Shaver, Gvineria, and Beck, 2008.

[5] The concept of perspectives was first introduced as part of a predecessor portfolio tool called DynaRank (Hillestad and Davis, 1998).

[6] This is in contrast to looking for robustness across perspectives. The danger in any filtering analysis is that good options might be eliminated prematurely, perhaps because of taking too seriously criteria that will later be relaxed.

Table 4.3
Illustrative Composite Options

Composite Option	Description

JTAC Package + Improved C2	JTAC capabilities are improved with
	  –anti-jamming
	  –ability to digitally share JTAC position and aircraft sensor point of interest
	  –ability to mark digital displays with "John Madden" aelestration
	  –ability to see full-motion video from targeting pod.
	C2 for CAS is streamlined with greater ASOC autonomy.

JTAC Plus Package + Improved C2	JTAC capabilities are improved as above. JFOs and more JTACs are added to the force structure. C2 for CAS is streamlined with greater ASOC autonomy.

Situational Awareness Package + Improved C2	JTAC capabilities are improved with
	  –anti-jamming
	  –ability to digitally share JTAC position and aircraft sensor point of interest.
	More UAVs are added.
	Total Blue Force Tracking is enabled.
	C2 for CAS is streamlined with greater ASOC autonomy.

Kill Package + Improved C2	JTAC capabilities are improved with
	  –anti-jamming
	  –ability to digitally share JTAC position and aircraft sensor point of interest
	  –standoff precision munitions are made available
	  –dial-a-yield munitions are made available.
	C2 for CAS is streamlined with greater ASOC autonomy.

CHAPTER FIVE

Portfolio Analysis of CAS Capability Options

The Portfolio-Analysis Structure

Given a set of options to be compared, as developed in Chapter Four, we need a framework within which to do it. Figures 5.1 and 5.2 show schematically the structure within which we evaluate options. At the top of Figure 5.1, we see that the criteria for evaluation include (1) warfighting effectiveness for supporting operations in four test-set CAS scenarios labeled Maneuver A, Maneuver B, Stabilization A, and Stabilization B; (2) the value for "other" missions; and (3) option risks.

As indicated in Figure 5.2, we decompose effectiveness for CAS missions into two levels of detail. Results can be viewed either at the Summary Level (Level 1) or, by zooming, at Levels 2 or 3. Level 3's structure corresponds to the decomposition of the mission that we described in Chapter Three when describing our set of capability models. Thus, the approach is integrated from a conceptual and analytical perspective. This would not be possible if the model used for analysis were an off-the-shelf detailed model developed for, say, training, mis-

Figure 5.1
Top-Level Portfolio Structure

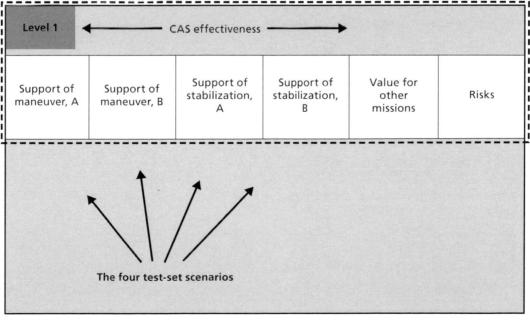

Figure 5.2
Second- and Third-Level Structures

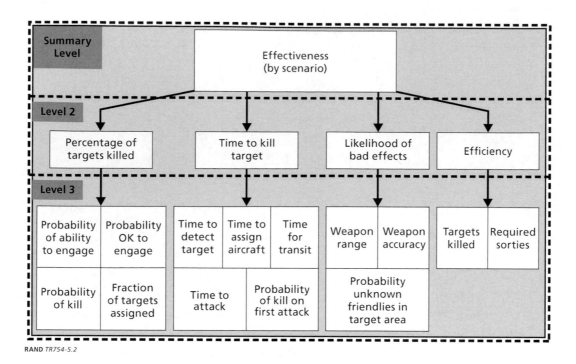

RAND *TR754-5.2*

sion planning, or engineering evaluation of weapon systems. We can also decompose "Value for Other Missions" and "Risks," but shall not discuss that further in this report.[1]

Illustrative Effectiveness Displays

Figure 5.3 shifts from the schematic illustrations to screenshots of the actual Portfolio Analysis Tool (PAT) (Davis and Dreyer, 2009), which is based on the familiar Microsoft Excel spreadsheet. This figure shows the summary level of results, although it excludes some information discussed later.[2] At the top of the display are various control panels, which can be used to change (1) the display's color coding and information, (2) the scoring method used to calculate an overall effectiveness from top-level components, (3) the "perspective" used in calculating that overall effectiveness, (4) level of detail for date entry, (5) the type of costs used for calculating cost-effectiveness, and (6) the total discount rate used in such calculations (inflation plus the "real" discount rate). These are best described elsewhere, but it should be understood that the analyst has great flexibility in using PAT.

Looking now to the "scorecard" part of Figure 5.3, we see that options appear in rows, whereas the measures discussed above are in columns. Colors and letters indicate the relative goodness of the options.[3] The 1s and 0s are the relative weights assumed for the various measures, which can be readily changed.

In this first example, the options shown are some of the many building-block options discussed in Chapter Four. Note that none of them do very well: Turning "reds" to "oranges" is not exactly success. The reason for this result is that the building-block options address only some contributors to mission effectiveness. The composite options discussed in Chapter Four

Figure 5.3
Illustrative PAT Display (notional data)

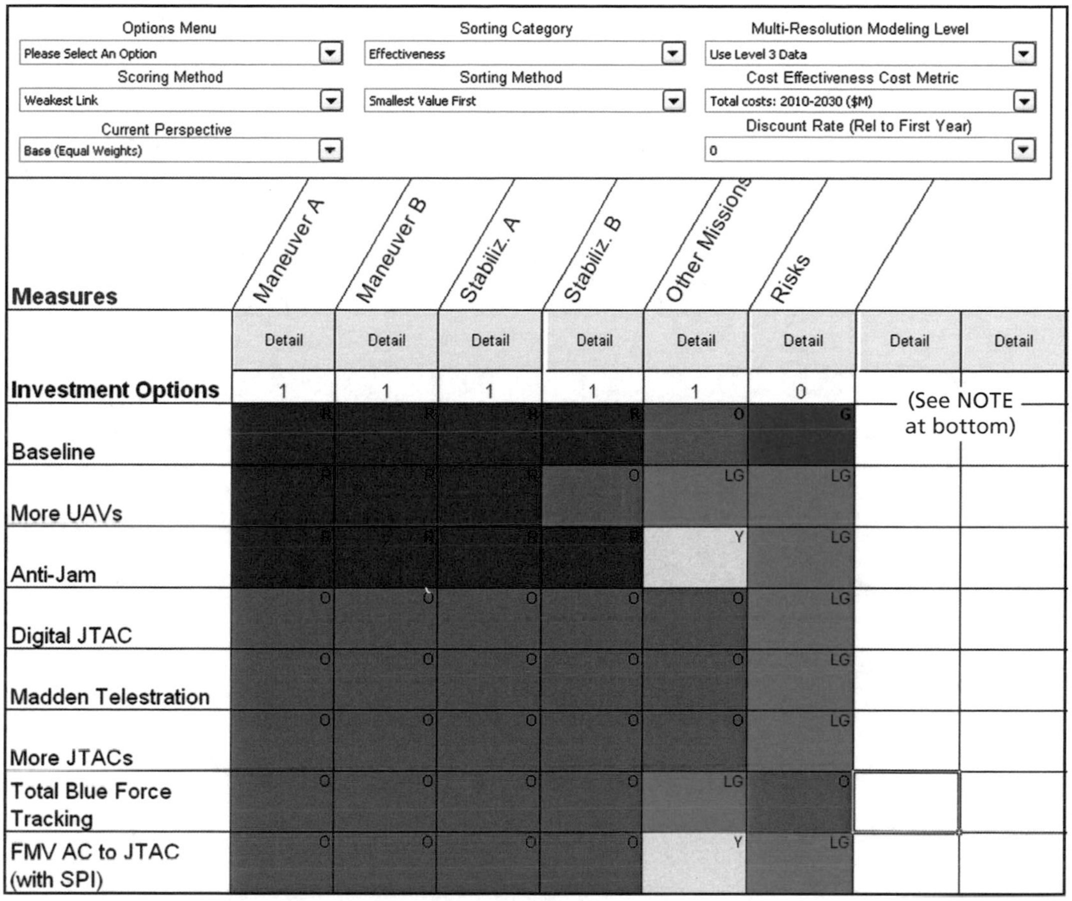

Measures	Maneuver A	Maneuver B	Stabiliz. A	Stabiliz. B	Other Missions	Risks		
	Detail	Detail	Detail	Detail	Detail	Detail	Detail	Detail
Investment Options	1	1	1	1	1	0	(See NOTE at bottom)	
Baseline	R	R	R	R	O	G		
More UAVs	R	R	R	O	LG	LG		
Anti-Jam	R	R	R	R	Y	LG		
Digital JTAC	O	O	O	O	O	LG		
Madden Telestration	O	O	O	O	O	LG		
More JTACs	O	O	O	O	O	LG		
Total Blue Force Tracking	O	O	O	O	LG	O		
FMV AC to JTAC (with SPI)	O	O	O	O	Y	LG		

NOTE: Many building-block options omitted for simplicity of graphic.
RAND *TR754-5.3*

were intended to be more suitable alternatives, because each combines improvement features so as to improve overall mission performance, not just one component.

Figure 5.4 shows a summary scorecard for the four composite options mentioned in Chapter Four and omits extraneous material, such as the control panels. Here, we see that all of the composite options improve results significantly, but they have different consequences for the different test cases. For example, the "JTAC Package + C2" composite option improves results significantly in maneuver cases, but does less well in stabilization cases, where enemy attacks can occur anywhere within very large areas. The larger JTAC package helps with that problem. The "Kill Package + C2" composite option does not improve matters very much, even though proponents of better weapons might have expected it to and could probably find special scenarios in which it would. The "Situational Awareness + C2 Package" option does the best of all the options, particularly in the relatively easier stabilization case (Stabilization A). It also has benefits for other missions, which might be assessed with other models or inputted by expert judgment.

To better understand the reasons for these results (although they are fictional, based on unclassified and merely notional "data"), we can zoom on the column for Maneuver A, obtain-

Figure 5.4
Summary Display with Composite Options (notional data)

Measures	Maneuver A	Maneuver B	Stabiliz. A	Stabiliz. B	Other Missions	Risks
	Detail	Detail	Detail	Detail	Detail	Detail
Investment Options	1	1	1	1	1	0
Baseline	R	R	R	R	O	G
JTAC Package + C2	Y	Y	O	O	O	LG
Large JTAC and C2 Package	Y	Y	Y	Y	O	LG
Sit. Aware. + C2 Package	Y	Y	LG	Y	LG	Y
Kill Package + C2	O	O	O	O	Y	Y

RAND *TR754-5.4*

ing Figure 5.5. That is, the output (last column of Figure 5.5) is the Maneuver A column in Figure 5.4. Such zooming allows the viewer—even a senior officer or official with time for only a few such questions—to "see" the basis of the higher-level score. Such zooming might be a spot check of the staff's work, a pursuit of more detail on something of particular interest, or an experiment to ensure understanding of how the analysis was accomplished and on what it

Figure 5.5
Zoom (Drill Down) on Support-Maneuver Case A (notional data)

Level 1 Measure	Maneuver A						
Level 2 Measure	Perc. of Targets Killed	Time to Kill Target Score	Likelihood of Bad Effects Score	Efficiency	Warning		
Investment Option							Maneuver A Score
Baseline	O	R	Y	O			R
JTAC Package + C2	LG	Y	LG	G			Y
Large JTAC and C2 Package	LG	Y	LG	G			Y
Sit. Aware. + C2 Package	G	Y	G	G			Y
Kill Package + C2	Y	O	LG	Y			O

RAND *TR754-5.5*

depends. The combining rules for deciding on the higher-level score can be rather complicated, but in most cases, the visual display is adequate explanation.

In this case, the limiting factors in the baseline assessment are the poor-to-mediocre results (red or orange) for killing targets quickly and the potential for bad effects, such as fratricide, collateral damage, and the strategic consequences thereof. In this example, our scoring simply takes the worst of the component scores—a common managerial approach when trying to assure mission effectiveness. To be sure, we could make the scores come out better by using a method such as simple averaging (a green and a red would average to a yellow), or a slightly more sophisticated approach, such as linear weighted sums, but to do so would be to hide problems. Consider the issue of timeliness. As discussed earlier, the real measure of effectiveness is not whether CAS kills targets eventually, but whether it kills (or otherwise suppresses) targets quickly enough—e.g., before a friendly ground-force unit is overrun or before those preparing or exiting from an IED emplacement escape. How fast is "fast enough" is a judgment that should be based on effects on the ground in actual operations and is very situation-dependent. As a point of comparison, Army artillery can sometimes put weapons on target within about five minutes or so of a call for fire—something that has historically been quite important. Those who argue that the Army does not need artillery because of airpower must confront the question of how quickly air power can respond and whether that is good enough. This said, Army units are increasingly being deployed to distant locations with minimal artillery, especially early in campaigns.

To see in more detail what is at issue in Figure 5.5, let us drill down again to obtain Figure 5.6. Here we see the components contributing to the time to kill targets. We see that in the baseline the primary problem has to do with the "Attack Time," the time an aircraft takes to execute the attack once it has reached the target area. This includes problems related to communications, delays in establishing whether the target can be safely attacked, and the potential need to reattack if the first attack fails. The composite options that add additional JTAC capability improve this situation. Nonetheless, the overall score never rises above "yellow," indicating marginal. The reason for this is that, while the component elements may seem reasonably

Figure 5.6
Zoom (Drill Down) on Time to Kill Target (notional data)

Level 2 Measure	Time to Kill Target Score						
Scoring Method	Thresholds						
Level 3 Measure	Detection Delay	Assignment Delay	Transit Time to Target	Attack Time	Prob (First-Time Kill)	Calc. Time to Kill Target	
Investment Option							Target Score Score
Baseline	LG	O	O	R	R	R	R
JTAC Package + C2	LG	LG	O	Y	G	Y	Y
Large JTAC and C2 Package	LG	LG	O	LG	G	Y	Y
Sit. Aware. + C2 Package	G	LG	O	Y	G	Y	Y
Kill Package + C2	LG	LG	O	O	O	O	O

good on an absolute basis (e.g., relatively quick detection, transit time, and so on), the overall time to attack is a sum of various delays. A *sum* of times, none of which seems long, can be too long for mission effectiveness. The transit time for the aircraft to reach the target is very difficult to improve on, so one would conclude that there are limits to what can be accomplished with CAS. In particular, CAS will seldom be as fast as well-placed artillery, although it may be more accurate.[4]

As a side note, the reader might argue that the explanation would be clearer if the criteria for the component times were tightened. That is easily accomplished, but not shown here. The results for the JTAC + C2 package, for example, might be green, yellow, light green, and yellow, and the overall yellow result might be more intuitive.

A Modern Depiction of Cost-Effectiveness Considerations

Cost-Effectiveness Landscapes

Another crucial element of portfolio analysis is related to economics: How do options compare when their costs are considered? RAND's approach to such matters is different from what has classically been taught in many business schools and operations research courses. In particular, we recommend against conducting analysis at only one budget level (i.e., reporting cost-effectiveness ratios), for several reasons.[5] First, decisionmakers need to see results as a function of cost to know how much additional funding would help or how badly having fewer funds would hurt. Second, cost numbers and budgets must both be viewed with suspicion. It is better to look at what we call *cost-effectiveness landscapes*. The value of doing so can be illustrated with some examples. One option might fall quite flat if its costs turn out to be 20 percent higher and no additional money is forthcoming. Conversely, one option might look better in an equal-cost analysis but have little upside potential, whereas another option might have substantially greater benefit with plausible increases of budget.

Figure 5.7 shows one cost-effectiveness landscape, again using made-up data. Each point in the diagram represents one of the options. When using PAT, the analyst can see the option names by "mousing" over a given point. We have annotated some of the points explicitly. The dashed line indicates the "efficient frontier" (i.e., the boundary beneath which the most cost-effective options sit).[6] In this depiction, the baseline option has effectiveness 0, so the chart is showing relative improvement of effectiveness for the various enhancements over the baseline.

Figure 5.7 shows that the Total Blue Force Tracking option fares poorly in this regard. It is roughly twice as expensive and yet only about as effective as the "JTAC and C2" option. If effectiveness is sufficiently important, then the "Situational Awareness and C2" option is best, but at a high price. It is significantly more effective than the "Large JTAC and C2" option.

The Importance of Making "Perspectives" Explicit

Unfortunately, such calculations of "overall effectiveness" are potential swindles because they suppress what can be *major* differences in judgment about the relative importance of the multiple criteria on which the options should be assessed, and how the criteria interact. Figure 5.7's cost-effectiveness landscape is for a particular perspective, one in which the various test-case scenarios (Maneuver A, Maneuver B, Stabilization A, Stabilization B) and the measure of "Value for Other Missions" are all valued equally (see label at bottom right). Another perspective, however, might be that permissive stabilization missions are far more important in the

Figure 5.7
Illustrative Cost-Effectiveness Landscape (notional data)

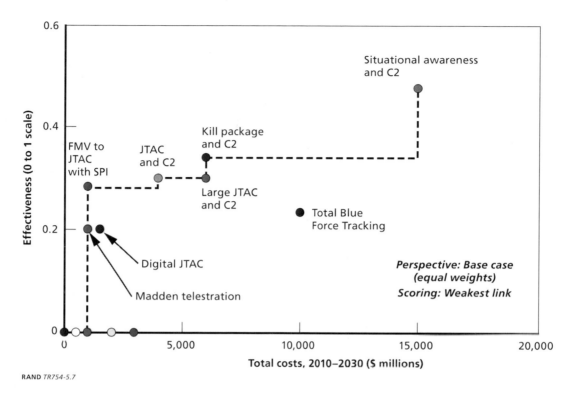

current and foreseeable world. Many leaders, including Secretary of Defense Robert Gates, have argued that U.S. planning has been imbalanced, focusing too much on hypothetical future wars and too little on the types of wars that we have been and are today fighting (Gates, 2009).

Yet another perspective, analogous to something mentioned earlier, would be to evaluate an option as poor if it were poor by *any* of several criteria. That is the perspective favored by those seeking robust capabilities.

As a separate matter, the assessments by the various criteria (e.g., effectiveness in one of the planning scenarios) may also be quite uncertain—due either to details of assumption (e.g., the precise capability of shoulder-fired SAMs) or to "soft" considerations, such as a reasoned guess about how well collateral damage can be avoided or how much trouble such damage will cause. In other work, we have defined what we call *extended perspectives* that vary by virtue of both perspective and assumptions (Davis, Shaver, and Beck, 2008).

So, how do we deal with the complications of multiple perspectives? More sophisticated analysis is possible (Davis, Shaver, Gvineria, and Beck, 2008), but the highest payoff is probably moving to charts, such as Figure 5.8, which highlight how the relationships among options change with perspective. Upon inspection, we see that the goodness and adequacy of the "Large JTAC and C2" option look much better in the perspective that emphasizes permissive stabilization (the bottom right panel).

In a real study, such a display might only be a starting point for competition. Those favoring one or another option would find ways to improve effectiveness, reduce cost, or both. Perhaps they could persuade evaluators to use different assumptions in evaluating effectiveness

Figure 5.8
Cost-Effectiveness Landscapes for Different Perspectives

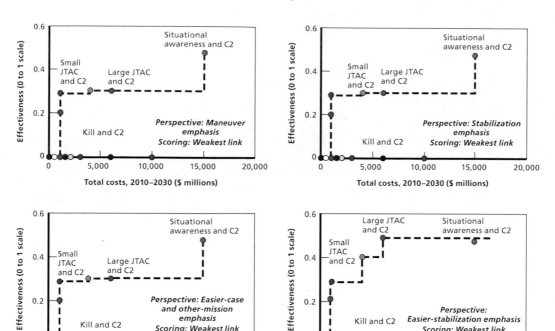

RAND *TR754-5.8*

in the different test-case scenarios. Iteration would occur as the competition unfolded. At the end, however, such a display—expressing the staff's best sense of what can be accomplished with different options—provides decisionmakers with a great deal of information. Staff might choose a perspective and look for the knees in the curve. Or they might look at results and conclude that the budget must be increased for the activities in question, because to obtain acceptable levels of effectiveness requires doing so. Yet another possibility is that they will ask whether the criteria used for assessments are too stringent. They might argue that the baseline for CAS capability is actually rather good in absolute terms, rather than as dire as suggested in Figure 5.4. Yes, timeliness is important, but should it dominate the assessment? How much timeliness is essential? The result could be a rapid reanalysis using a scoring approach more like the familiar method of linear weighted sums. Using PAT, such reanalysis can be done in minutes (most of which would be used to format and annotate graphics).

Additional Comments About Portfolio Analysis

Discussion of our portfolio methods could be much more extensive, but what matters most is conveying a sense of philosophy. Investment decisions should be based on five factors:

- strategic priority as established by decisionmakers
- desire for portfolio "balance" (across cases, functions, etc.), as discussed by Secretary Gates
- practical significance to warfighters; significance to both effectiveness and efficiency

- ability to meaningfully affect problems (invest where it will make a difference, not just by the importance of the problem)
- cost and cost-effectiveness considerations.

This list may seem unexceptionable, but it is in conflict with the common tendency to merely follow the priority of top officials expressed out of context (as in a policy document or a public speech). We urge paying attention to all of these factors, which is actually no more than assuming that top decisionmakers are much more intelligent and reasonable than simple priority lists might suggest. No matter how important some problem is, investing in an expensive option associated with the problem is inappropriate if it would be unlikely to accomplish anything significant.[7] Nor should high priorities necessarily be "fully funded" before next-priority items are partly funded. In practice, programs to address high priorities usually have high-leverage components and other components that are either more dubious or less cost-effective. They should perhaps not be funded if major progress can be made on an important but lesser priority for the same investment.

In our experience, this proves wise, effective, and helpful to decisionmakers. However, it implies changes of process and, e.g., the formats of discussion and measures of values used.

A second element of portfolio philosophy is recognizing that improving capabilities often requires managerial, procedural, or C2 changes, rather than changes only in what is "bought." Omitting such matters from options and analysis is indefensible to anyone interested in results. A management "trick" that can sometimes be used is to create "programs" to assure that the needed changes are accomplished. This is also useful because there often are expenses involved (retraining, reorganization, reworking of software, etc.). They may be small-dollar costs, but—unless programmed—may preclude actions.

Notes

[1] The item for "likelihood of bad effects" here could be interpreted to include concern about attrition, but more generally we would have included it explicitly. In the analysis, it was assumed that an aircraft would use a standoff weapon if available and needed, and not attack a defended target for which a suitable standoff weapon was not available.

[2] The actual spreadsheet display includes, to the right of the material shown, information on, e.g., costs, net effectiveness, and relative cost-effectiveness. Those are discussed in a later section. It is important for decisionmakers to understand the multicriteria assessments of effectiveness before getting into cost-benefit comparisons.

[3] Red (R), orange (O), yellow (Y), light green (LG), and green (G) correspond to what would be referred to qualitatively as very bad, bad, marginal, good, and very good. What these correspond to in the physical world depends on the specific problem and analytic choices. If one seeks an upbeat presentation for general purposes, the evaluations might be considerably easier than if the purpose is to develop *robust* capabilities, in which case it is important to ferret out and highlight problems that could affect overall mission performance.

[4] Technically, one solution would be to procure numerous armed UAVs so that there would be enough to ensure rapid response. Another solution would be to acquire standoff weapons with great range and high speeds. Such options could be quite expensive, of course, and would have limitations related to accuracy, collateral damage constraints, and other factors. In some cases, of course, the stack of CAS aircraft could be placed much closer to targets.

[5] Actually, cost-effectiveness is problematic even for a single budget level if one must solve the "knapsack" problem. Cost-effectiveness as a metric works when one can allocate a small increment of budget and actually

buy something with that small increment. In that case, it is "marginal analysis," which does optimally select in order of most cost-effective first, although with some potential logical artifacts.

[6]   Some readers may be more familiar with charts in which the efficient frontier is drawn as smoothly continuous. That is appropriate in theoretical economics. However, it makes assumptions about the existence of unplotted intermediate options that would define such a smooth curve. In our work, we consider only the options for which points are plotted.

[7]   An interesting example in recent years has involved SOF. Some critics have seemed to believe that such forces should be doubled or tripled in size, and have been disappointed when that didn't happen. However, those responsible for contemplating such changes argue that only relatively modest changes are feasible without severely undercutting the quality and effectiveness of the units: The people needed are inherently in short supply.

CHAPTER SIX

Conclusions and Recommendations for Next Steps

Operationalizing the Application to the CAS Mission Area

The work reported here was prototypical; it used highly notional "data" that was in no way intended to be realistic for a particular defense planning scenario. We did, however, attempt to capture the primary features of the CAS mission in our operational model and demonstrate that the model was both working properly and capturing what we understood from our research, including a field visit to a live-fire exercise and discussion with JTACs and CAS pilots.

A next step on this particular mission area would involve obtaining more realistic data and establishing a suitable set of test-case scenarios for serious analysis. The question then arises where better data might come from. Several possibilities exist, notably (1) operational experiences in the field, (2) exercises, (3) training, and (4) discussion with system engineers and others responsible for developing and refining CAS capabilities. The form of the data could include empirical reports, results from modeling and simulation, judgments obtained from highly structured interviews, and interpretations of other accounts.

In practice, obtaining such data is not so easy: These sources of data are seldom in the right form for investment analysis. It is not a matter of just changing formats or some such; rather, the data that have been collected do not answer the right questions. The good news, we believe, is that those using models and simulations for training, operations planning, and exercises could collect much more relevant information if the appropriate requests were made with sufficient clarity and rationale. We hope that our models and methodology will prove to be a mechanism for defining those requests.

Another problem is that the engineering-level people who often provide the options are at too deep a level to provide routinely the information needed for program analysis. Again, however, we hope that our work will facilitate communication. System engineers understand decomposition. Ours is merely a higher-level decomposition than they often work with.

Cost data constitute another major challenge. Some of the principal problems here are interoperability, scale, and cost-sharing (e.g., Who will pay what fraction of the cost for an initiative with value to multiple missions?).

Broadening Work to the "Enterprise Level": Making Assessments Across Capability Areas

The prototype work we have described focused on a particular mission area so as to better demonstrate concepts in a concrete manner. However, from the outset of the project, the intention has been to broaden—and indeed elevate—the work for analysis *across* missions and in a form suitable for enterprise-level discussion. The ultimate challenge includes developing methods for ensuring that good options supporting warfighter needs generally will compete effectively at the corporate Air Force level against options that have more visceral appeal in many respects (e.g., protecting the purchase of F-22s and JSFs).

Portfolio Balancing at the Enterprise Level

Some key elements of our approach will be portfolio balancing as a concept. Such balancing must be across missions, scenarios, and time horizons. The options must be highly aggregated packages, since enterprise-level decisionmaking should deal with those rather than the "piece parts."

Success in having good-for-warfighter options compete effectively will depend on good packaging and appropriate cross-cutting analysis. Enterprise-level officers (and civilian officials) will need to understand the implications of the packages and will need to have a good basis for believing that the options being considered have been well constructed and would indeed accomplish what is proposed. They should also be able to see that, in many cases, the true benefit of an option will not be adequately reflected by analysis for any single area.

Readers should have no illusions about how easy such enterprise-level work will be. In addition to the need for mission-level work in multiple domains, it will be necessary to find ways to clarify the nature of trade-off across those domains. We say "clarify the nature" because there will be no operations-research "solution" to the challenge. Such trade-offs are inherently in the realm of strategic decisionmaking and involve a mix of objective and subjective considerations (one reason for our emphasis on perspectives, as described in Chapter Five).

Identifying Other Management-and-Process Changes Needed

A fundamental problem in attempting to ensure that warfighter needs receive sufficient support in enterprise-level decisionmaking is that many such needs are inherently "low-level" in nature. Although we did not examine such issues in any depth in our study, it was clear to us that many of the CAS-improvement measures we learned about "should" be decided on at lower levels of the organization. That would be in keeping with the widely understood management concept of moving decisionmaking to the appropriate level.[1] Many of the failures to address warfighter needs could be avoided if funding on various "mundane" but important issues were decided at lower levels, closer to the operators.

This may seem like a platitude, but there are important subtleties. The procedure of providing pots of resources to lower-level commanders to use intelligently implies less money for top leaders to be allocating. In an era marked by money shortages for high-visibility acquisition programs and other such matters, the pressures to squeeze lower-level budgets to ameliorate top-level problems are high.

Another problem is even more subtle, but fundamental. Hierarchical organizations, of which the Air Force is merely one among many, are prone to the *tyranny of the prioritization list*. When the Secretary of Defense, Combatant Commanders, and Chief of Staff express their priorities, those who must allocate resources may take the approach of "racking and stacking" proposals for funding and then move downward from the top priority, funding the proposals until money is exhausted. The terminologies used here include "funding the above-the-line items."

There is considerable intuitive appeal to such an approach, which is ubiquitous. However, it is at odds with system thinking and portfolio management. It is not appropriate to invest heavily in an option merely because it purports to address a high-priority issue. What if it is ineffective? What if it is extremely expensive and not very effective, whereas a smaller investment would make a big difference in something that is also very important but not of a character suitable for a chief-of-organization's priority list? What if a proposal addressing a high-priority issue could be decomposed into "tranches," with the first tranche having a high impact at reasonable expense and with subsequent tranches having much less benefit per unit of expenditure? Shouldn't those subsequent tranches compete with investments to improve important capabilities not at the top of the priority list?

Astute readers will perhaps say "Well, of course, but such considerations are taken into account implicitly as the options are created in the first place: Some important low-level investments are slipped into packages named in terms of high-priority matters, and the options are scrutinized so as not to go beyond the realm of efficient investment (i.e., to exclude what we referred to above as second- and third-tranche components)." If only this were consistently true. But it is not. We believe that an important aspect of the next phase of work should be examining Air Force processes so that it is more true.

Relating Work to Other Air Force Efforts on Risk Management

The Air Force CRRA process is the Air Force capabilities planning process led by the Directorate of Operational Capability Requirements (HQ USAF A5XC) and is designed to provide Air Force leaders an operational assessment of current and future AF capabilities to meet joint warfighting requirements in order to influence strategic planning and investment strategies. The CRRA process relies heavily on subject-matter experts (SMEs) and mission process sequence models to provide detailed assessments of probability of overall mission success within a range of scenarios and time frames. A separate group of SMEs evaluates the consequences of mission failure, and the two are combined to assess the risks and risk drivers.

The CRRA process and the capability modeling process we have described in this report have in common the decomposition of missions, the use of SMEs, and the use of scenarios (although we have suggested a broader range of scenarios and scenario variables). The major differences are the use of multiple measures of effectiveness, cost and effectiveness analysis of composite options, and portfolio analysis in the RAND approach, as described in this report.

We believe that there would be considerable synergy in combining the two approaches to provide the Air Force with a major capabilities-based portfolio-analysis tool for corporate program development, assessment, and justification. As part of the next stage of this project, we will also work with HQ USAF A5XC to identify the possible approaches to integrating these methods.

Note

[1]   See, e.g., Simons, 2005.

The CASEM Model

Introduction

The CASEM simulates aircraft prosecuting a series of ground targets. Aircraft are launched at a regular basis, providing a constant amount of available air power, while targets of different types appear randomly. Aircraft are assigned to targets based on the probability of successful engagement, but can be reassigned to higher-priority targets as needed. Targets must be engaged within a certain amount of time (randomly chosen for each target) to be effective. This time may correspond to, e.g., how quickly enemy ground forces need to be engaged lest they have serious deleterious effects on the ground battle, or how quickly enemy forces (even small groups of people laying ground mines) must be attacked before they will disappear.

Failures may occur because of shortages of aircraft relative to the number of targets, limitations of ISR platforms, the attack aircraft not finding the target, communication problems, anti-air defenses preventing the aircraft from getting close enough to release a weapon, the target being too close to either friendly forces or other collateral damage concerns for the weapons available on the aircraft, or—as indicated above—response being too slow.

As discussed in Chapter Three, CASEM and our simplified model both focus on relatively abstract measures of flexible, adaptive capabilities, rather than on, say, simulated results of simulations rich with details of ground-force and air-force operations and their interactions. That is, despite our being adamant about the need for the CAS measures to relate to addressing ground commanders' needs, we conclude that this can be accomplished by using what appears to be a more standard target-servicing approach—so long as the targets of our model are challenging enough in terms of number, density in time, diversity, and—importantly—the times within which they must be attacked (e.g., before they cause undue difficulty for friendly ground-force operations or before they disappear in hit-and-run tactics).

Model Variables

Scenario Parameters

$D_{B,S}$	=	Distance from the air base from which the aircraft launch to the "stack," the point in theater from which aircraft begin CAS attacks.
D_{max}	=	Maximum distance from the stack to a target.
$Pr_{Acq,ISR}$	=	Probability that ISR assets pick up the target when it appears.

$\Pr_{Acq,CAS}$ =   Probability that the CAS aircraft sees the target (assuming the ISR assets see the target first).

T_{Sc} =   The total time that the scenario runs.

$T_{G,D}$ =   Time between the appearance of the target and its detection by ISR assets.

$T_{D,A}$ =   Time between target detection and attack aircraft being assigned to the target.

T_{Att} =   Time between subsequent attacks on a given target by the aircraft assigned to a particular attack sortie.

$\Pr_{Comm}$ =   Probability that communications are operating during the attack. If communications fail during a single attack, the attack is called off and another attack is attempted. The time lost to the attack is the same as if the attack had occurred and failed.

Aircraft Parameters

R_S =   Sortie rate: the rate at which aircraft take off from the airbase.

AC_A =   Number of aircraft per attack. The effect of multiple aircraft in an attack for this model is as a multiplier to the number of weapons available to prosecute ground targets. It also affects the number of separate targets that can be engaged in a target-rich environment.

AC_V =   The cruise velocity of the aircraft.

AC_E =   The aircraft's mission time (includes transit time and time on station).

$\Pr_{BDA}$ =   The probability that the aircraft performs correct BDA. This probability covers both the probability of correctly identifying a destroyed target as well as the probability of detecting that a target requires additional attacks.

Weapon Parameters (up to two weapons can be defined per aircraft)

N_{W_I} =   Number of weapons of type I on each aircraft.

$Pk_{I,J}$ =   Single-shot probability of kill of weapon type I against target class J.

D_{W_I} =   Range of weapon type I (to be compared with standoff range necessitated by defenses).

$D_{CD_I/FF_I/NC_I}$ =   The minimum distance that a collateral damage concern (CD) or friendly forces (FF) or noncombatant (NC) can be from a target in order to prohibit a weapon of type I from being released.

Target Parameters (up to two classes of targets can be defined per scenario)

R_J =   Target class J generation rate (times between subsequent targets of class J are modeled using an exponential distribution with average $1/R_J$).

$Val_{minJ/maxJ}$ =   Minimum and maximum values for targets of class J. Each target of class J is assigned a value (importance) from a uniform distribution between Val_{minJ} and Val_{maxJ}.

$KO_{\text{min}J/\text{max}J}$ =   Minimum and maximum keepout range for targets of class J. Each target of class J is assigned a keepout range from a uniform distribution between $KO_{\text{min}J}$ and $KO_{\text{max}J}$. If a weapon's range is less than the keepout range of the target, then an aircraft is not allowed to use that weapon, as the keepout range signfies the range at which air defenses surrounding a target could engage the attacking aircraft.

$DT_{\text{min}J/\text{max}J}$ =   Minimum and maximum dwell time for targets of class J. Each target of class J is assigned a dwell time from a uniform distribution between $DT_{\text{min}J}$ and $DT_{\text{max}J}$. The dwell time denotes the amount of time after a target is generated that it can be engaged. Afterward, it is assumed that the target has escaped.

$CD_{\text{min}J/\text{max}J}$
$FF_{\text{min}J/\text{max}J}$
$NC_{\text{min}J/\text{max}J}$ =   Minimum and maximum collateral damage/friendly forces/noncombatant distances for targets of class J. Each target of class J is assigned these three distances from a uniform distribution between the appropriate minimum and maximum values. If any of these distances are within the associated minimum distances of both weapons on the aircraft, the aircraft cannot engage the target.

Model Description

After creating queues of targets and aircraft, the model assigns aircraft to available targets and models each engagement. In this context, *available* means that the target has been spotted by ISR assets; is not close enough to other objects to cause any friendly force, noncombatant, or collateral damage; and does not have a keepout range larger than the range of at least one of the weapons on the aircraft. The model iteratively selects the earliest available target and assigns an aircraft to each one (if possible). Of the aircraft that are currently in the air, the model assigns an aircraft to the target that maximizes the increase in the expected total value of targets destroyed. For each aircraft in the air, including ones already assigned to targets, the model determines the number of rounds of attacks the aircraft could do before running out of weapons and/or fuel, and determines the probability of successfully engaging the target. This probability of engagement is multiplied by the importance of the target to get an expected value of the attack. If there is a tie between two or more aircraft, the model assigns the aircraft that has the least time before it has to return to base. For aircraft already assigned targets, the expected value of killing the new target is reduced by the expected value of killing the already assigned target. If an aircraft is pulled off of engaging a target, it is possible that another aircraft will engage the target instead, if one is available.

The engagement itself is quite simple. If the aircraft sees the target and the communications with the JTAC on the ground is operational, then the aircraft releases the best available weapon at the target (that is, the one with the highest P_k that can be fired while avoiding concerns of safety to noncombatants, etc., and the aircraft itself). If the aircraft engages the target successfully and performs correct BDA, then the attack is considered a success. If the aircraft engages the target successfully, but makes an incorrect BDA, then the aircraft will attack the target again (if possible). If the aircraft is not successful in engaging the target and makes a correct BDA, then the aircraft will attack the target again (if possible). If the aircraft is not successful in engaging the target and also makes an incorrect BDA, then the aircraft will consider that it has done its job, even though the target is not destroyed. After an engagement ends (either successfully or not), an aircraft is available again to attack other targets until it runs out of fuel and/or weapons.

The fraction of targets that will be successfully attacked depends on many factors:

- Target visibility by ISR and the aircraft: The visibility of the target by the attack aircraft is conditioned by an ISR asset seeing it as well, so the probability this occurs is

$$\text{Pr}_{Acq,ISR} \times \text{Pr}_{Acq,AC}.$$

- Keepout (and other) range concerns: The probability that keepout range concerns will stop an attack from a weapon of type I from occurring is[1]

$$\min\left(\max\left(0, \frac{D_{W_I} - KO_{\min}}{KO_{\max} - KO_{\min}}\right), 1\right).$$

- The probabilities that collateral damage or friendly forces or noncombatants will stop an attack have an identical format. As these probabilities are independent, the probability that any one of these issues will prevent an attack is just one minus the probability that none of these problems occur.
- Availability of aircraft. Each aircraft will be able to engage targets for time equal to its endurance time. However, attacks are in groups. Thus, the steady-state number of aircraft groups available at any time is roughly $AC_{END}(R_S/AC_A)$, where

$$AC_{END} = AC_E - 2x\frac{D_{B,S}}{AC_V}.$$

- Time to engage targets. Let DT be the dwell time of a particular target of class J. The time that the aircraft has to engage the target with a weapon of type I, DT_I, is reduced by the amount of time it takes to find the target, assign it to an aircraft, and fly the aircraft a distance $D_{AC,Tgt}$ to within range of the target:

$$DT_I = DT - T_{G,D} - T_{D,A} - \frac{\max(0, D_{AC,Tgt} - D_{W_I})}{V_{AC}}.$$

The maximum number of attacks is

$$N_{A_I} = \min\left[\left(\frac{DT_I}{T_{Att}}\right), N_{W_I}\right],$$

and so the probability of an attack with weapons of type I succeeding and having correct BDA (assuming all previous things have gone correctly) is

$$(Pk_{I,J} \times \text{Pr}_{Comm} \times \text{Pr}_{BDA}) \sum_{k=0}^{N_{A_I}-1}\left\{1 - \text{Pr}_{Comm} \times \left[Pk_{I,J} \times \text{Pr}_{BDA} + (1 - Pk_{I,J}) \times (1 - \text{Pr}_{BDA})\right]\right\}^k.$$

Inputs and Outputs

Model Inputs

- System-level parameters
  - Distance, base to stack
  - Size of CAS stack area of responsibility
  - Pr(target acquired by ISR)
  - Pr(target acquired by aircraft, given acquisition by ISR)
  - Detection delay
  - Assignment delay
  - Time per attack
  - Pr(communications throughout attack)
- Aircraft parameters
  - Sortie rate
  - Aircraft per attack
  - Speed
  - Endurance
  - Pr(correct BDA)
- Weapon parameters (multiple)
  - Number per aircraft
  - Single-shot probability of kill
  - Weapon range
  - Minimum avoidance-distance to avoid collateral damage, fratricide
- Target parameters (multiple)
  - Generation rate
  - Value (treated as stochastic variable)
  - Air defenses (keepout range; treated as stochastic variable)
  - Dwell time (treated as stochastic variable)
  - Distance from friendly forces, noncombatants, and other to-be-avoided objects

---

Model Outputs

- Timely responsiveness to ground-force requests
- Targets destroyed, with correct BDA
- Targets destroyed per sortie
- Targets destroyed, with incorrect BDA
- Failures (target not destroyed in timely manner), with breakdown:
    - Target disappears before attack
    - Target not acquired by ISR
    - Target acquired by ISR, not acquired by aircraft
    - Target acquired, no aircraft assigned
    - Target not engaged (too close to collateral damage sites)
    - Target not engaged (too close to friendly forces)
    - Target not engaged (too close to noncombatants)
    - Target not engaged (too close to air defenses)
    - Target engaged, aircraft reassigned
    - Target engaged, not killed
    - Target engaged, wrong BDA (false positive)

Note

[1]   To avoid divisions by zero in odd cases, users should add a small number (e.g., 0.0001) to denominators such as this.

A Motivated Metamodel Connected to CASEM

A "motivated metamodel" is usually a simple formula model obtained by drawing on an intuitive understanding of the problem to postulate an approximate functional form and then testing and calibrating that form by statistical analysis of outputs from a more detailed model—outputs resulting from an experimental design that generates an appropriate diversity of cases for the detailed model. The result can have much greater explanatory capability than standard statistical metamodels obtained by ordinary regression analysis (linear sums, perhaps with some interaction terms).[1]

The appropriate way to do motivated metamodeling is to include explicit correction terms in the postulated form. If, for example, one imagined that the outcome function F should be roughly proportional to the product of A, B, and C (perhaps because output was something like probability of mission success, dependent on each of several component parts of the mission being successful), then the postulated form might be:

$$F = C_1 ABC(1 + C_2 A + C_3 B + C_4 C) + C_5.$$

The usual methods of linear regression would then be used to find the various coefficients (C_1, C_2, . . .) that are a best fit to the output data of the detailed model. However, instead of treating the variables A, B, and C as the linear objects of regression, one would use ABC, A^2BC, AB^2C, and ABC^2. If the statistical analysis concludes that all but C_1 are ignorably small, then one has verified the usefulness of the postulated model and found a calibration factor (C_1) that allows it to work well. On the other hand, if the other coefficients are not ignorably small, then it may be necessary to reconsider the postulated form.

For the prototype work, it turned out that the simple multiplicative model shown in Figure 3.3 was a rather good fit to the results from CASEM, and we did not actually go through a full statistical analysis. In a fuller analysis with more realistic data, that might not be true, and correction terms might prove necessary.

Note

[1] The theory and rationale for motivated metamodeling are discussed in Davis and Bigelow, 2003.

Details of Portfolio-Analysis Structure

Table C.1 shows the data structure used for our illustrative portfolio analysis. It corresponds closely to discussion in the text but illustrates the format used to enter structuring information in PAT.

Table C.1
Structure for Portfolio Analysis

Level One Measure	Level Two Measure	Level Three Measure
Maneuver effectiveness 1A	Responsiveness to ground forces	Execution probability
		Delay time
		Standoff if needed?
		Capacity
		Calculated responsiveness
	Broad effectiveness	Execution probability
		Delay time
		Standoff if needed?
		Capacity
		Calculated responsiveness
	Minimal bad effects	
Efficiency, maneuver 1A	Efficiency	
Maneuver effectiveness 1B	Responsiveness to ground forces	Execution probability
		Delay time
		Standoff if needed?
		Capacity
		Calculated responsiveness
	Broad effectiveness	Execution probability
		Delay time
		Standoff if needed?
		Capacity
		Calculated responsiveness
	Minimal bad effects	
Efficiency, maneuver 1B	Efficiency	

Table C.1—continued

Level One Measure	Level Two Measure	Level Three Measure
Stabilization effectiveness 2A	Responsiveness to ground forces	Execution probability
		Delay time
		Standoff if needed?
		Capacity
		Calculated responsiveness
	Broad effectiveness	Execution probability
		Delay time
		Standoff if needed?
		Capacity
		Calculated responsiveness
	Minimal bad effects	
Stabilization efficiency, 2A	Efficiency	
Stabilization effectiveness, 2B	Responsiveness to ground forces	Execution probability
		Delay time
		Standoff if needed?
		Capacity
		Calculated responsiveness
	Broad effectiveness	Execution probability
		Delay time
		Standoff if needed?
		Capacity
		Calculated responsiveness
	Minimal bad effects	
Stabilization efficiency 2B	Efficiency	
Risks	Assessment risk	
	Technical risk	
	Program-schedule risk	
Value for other missions		

Bibliography

Bowie, Christopher J., Fred L. Frostic, Kevin N. Lewis, John Lund, David Ochmanek, and Philip Propper, *The New Calculus: Analyzing Airpower's Changing Role in Joint Theater Campaigns*, Santa Monica, Calif.: RAND Corporation, 1993. As of January 14, 2010:
http://www.rand.org/pubs/monograph_reports/MR149/

Davis, Paul K., *Effects-Based Operations (EBO): A Grand Challenge for the Analytical Community*, Santa Monica, Calif.: RAND Corporation, 2001. As of January 14, 2010:
http://www.rand.org/pubs/monograph_reports/MR1477/

———, *Analytic Architecture for Capabilities-Based Planning, Mission-System Analysis, and Transformation*, Santa Monica, Calif.: RAND Corporation, 2002. As of January 14, 2010:
http://www.rand.org/pubs/monograph_reports/MR1513/

Davis, Paul K., and James H. Bigelow, *Experiments in Multiresolution Modeling (MRM)*, Santa Monica, Calif.: RAND Corporation, 1998. As of January 14, 2010:
http://www.rand.org/pubs/monograph_reports/MR1004/

———, *Motivated Metamodels: Synthesis of Cause-Effect Reasoning and Statistical Metamodeling*, Santa Monica, Calif.: RAND Corporation, 2003. As of January 14, 2010:
http://www.rand.org/pubs/monograph_reports/MR1570/

Davis, Paul K., James Bonomo, Henry Willis, and Paul Dreyer, "Analytic Tools for Strategic Planning and Investment in the Missile Defense Agency," in *Proceedings of the 3rd Annual U.S. Missile Defense Conference*, 2005.

Davis, Paul K., and Paul Dreyer, *RAND's Portfolio Analysis Tool (PAT): Theory, Methods, and Reference Manual*, Santa Monica, Calif.: RAND Corporation, TR-756-OSD, 2009. As of January 14, 2010:
http://www.rand.org/pubs/technical_reports/TR756/

Davis, Paul K., and Amy Henninger, *Analysis, Analysis Practices, and Implications for Modeling and Simulation*, Santa Monica, Calif.: RAND Corporation, OP-176-OSD, 2007. As of January 14, 2010:
http://www.rand.org/pubs/occasional_papers/OP176/

Davis, Paul K., and Richard Hillestad, "Families of Models That Cross Levels of Resolution: Issues for Design, Calibration, and Management," in G.W. Evans, M. Mollaghasemi, E.C. Russell, and W.E. Biles, eds., *Proceedings of the 1993 Winter Simulation Conference*, New York: Association for Computing Machinery, 1993, pp. 1003–1012.

Davis, Paul K., and James P. Kahan, *Theory and Methods for Supporting High-Level Decisionmaking*, Santa Monica, Calif.: RAND Corporation, TR-422-AF, 2007. As of January 14, 2010:
http://www.rand.org/pubs/technical_reports/TR422/

Davis, Paul K., Russell D. Shaver, and Justin Beck, *Portfolio-Analysis Methods for Assessing Capability Options*, Santa Monica, Calif.: RAND Corporation, 2008. As of January 14, 2010:
http://www.rand.org/pubs/monographs/MG662/

Davis, Paul K., Russell D. Shaver, Gaga Gvineria, and Justin Beck, *Finding Candidate Options for Investment: From Building Blocks to Composite Options and Preliminary Screening*, Santa Monica, Calif.: RAND Corp, TR-501-OSD, 2008. As of January 14, 2010:
http://www.rand.org/pubs/technical_reports/TR501/

Department of Defense, Directive on Capability Portfolio Management, Directive 7045.20, September 25, 2008.

Deptula, Brigadier General David (USAF), *Effects-Based Operations: Change in the Nature of Warfare*, Arlington, Va.: Aerospace Education Foundation, 2001.

Deptula, Lieutenant General David (USAF), "Effects-Based Operations: A U.S. Commander's Perspective," *Air and Space Power Journal*, Spring 2006.

Dubik, James, "Effects-Based Decisions and Actions," *Military Review*, January–February 2003, pp. 33–36.

Elder, MG Robert J., "Effects-Based Operations in Operation Iraqi Freedom," briefing, 2006.

Gates, Robert, "A Balanced Strategy: Reprogramming the Pentagon for a New Age," *Foreign Policy*, Vol. 88, No. 1, January/February 2009.

Grant, Rebecca, "Armed Overwatch," *Air Force Magazine*, Vol. 91, No. 12, 2008.

Grossman, Elaine M., "Effects-Based Operations Under Fire: A Top Commander Acts to Defuse Military Angst on Combat Approach," *Inside the Pentagon*, April 20, 2006.

Hillestad, Richard, and Paul K. Davis, *Resource Allocation for the New Defense Strategy: The DynaRank Decision-Support System*, Santa Monica, Calif.: RAND Corporation, 1998. As of January 20, 2010: http://www.rand.org/pubs/monograph_reports/MR996/

Hughes, Wayne, ed., *Military Modeling*, 2nd ed., Alexandria, Va.: Military Operations Research Society, 1989.

Hura, Myron, Gary McLeod, Richard Mesic, Philip Sauer, Jody Jacobs, Daniel M. Norton, and Thomas Hamilton, *Enhancing Dynamic Command and Control of Air Operations Against Time Critical Targets*, Santa Monica, Calif.: RAND Corporation, 2000. As of January 14, 2010: http://www.rand.org/pubs/monograph_reports/MR1496/

Jacobs, Jody, Leland Joe, David Vaughan, Diana Dunham-Scott, Lewis Jamison, and Michael Webber, *Technologies and Tactics for Improved Air-Ground Effectiveness*, Santa Monica, Calif.: RAND Corporation, 2008, Not Available to the General Public.

Jacobs, Jody, David E. Johnson, Katherine Comanor, Lewis Jamison, Leland Joe, and David Vaughan, *Enhancing Fires and Maneuver Capability Through Greater Air-Ground Joint Interdependence*, Santa Monica, Calif.: RAND Corporation, 2009. As of January 20, 2010: http://www.rand.org/pubs/monographs/MG793/

Jacobs, Jody, Gary McLeod, and Eric V. Larson, *Enhancing the Integration of Special Operations and Conventional Air Operations: Focus on the Air-Surface Interface*, Santa Monica, Calif.: RAND Corporation, 2007, Not Available to the General Public.

Jobbagy, Zoltan, *Effects-Based Operations and the Problem of Thinking Beyond: A Critical Reflection*, The Netherlands: Centre for Strategic Studies, 2006.

Johnson, Stuart, Martin Libicki, and Gregory Treverton, *New Challenges, New Tools for Defense Decisionmaking*, Santa Monica, Calif.: RAND Corporation, 2003. As of January 14, 2010: http://www.rand.org/pubs/monograph_reports/MR1576/

Joint Chiefs of Staff, *Joint Tactics, Techniques, and Procedures for Close Air Support (CAS)*, Washington, D.C.: Department of Defense, 2003.

Joint Chiefs of Staff Instruction, *Joint Capabilities Integration Development System*, CJCSI 3170.01G, Washington, D.C., March 1, 2009.

Jones, James, and Robert Herslow, "The United States Air Force Approach to Capabilities-Based Planning and Programming (CBP&P), Parts 1 and 2," *Meeting Proceedings RTO-MP-SAS-055*, Paper 10,1, Neuilly-sur-Seine, France: RTO, 2005.

Kelley, Charles T., Jr., Paul K. Davis, Bruce W. Bennett, Elwyn Harris, Richard Hundley, Eric V. Larson, Richard Mesic, and Michael D. Miller, *Metrics for the Quadrennial Defense Review's Operational Goals*, Santa Monica, Calif.: RAND Corporation, DB-402-OSD, 2003. As of January 14, 2010: http://www.rand.org/pubs/documented_briefings/DB402/

Kirkpatrick, Charles E., *Joint Fires as They Were Meant to Be: V Corps and the 4th Air Support Operations Group During Operation Iraqi Freedom,* Fort Leavenworth, Kan.: Institute of Land Warfare, Association of the United States Army, Land Warfare papers, No. 48, 2004.

Lucas, Thomas, "The Stochastic Versus Deterministic Argument for Combat Simulations: Tales of When the Average Won't Do," *Military Operations Research Journal,* Vol. 5, No. 3, 2000, pp. 9–28.

Mattis, General James N., "USJFCOM Commander's Guidance for Effects-Based Operations," *Parameters,* Autumn 2008, pp. 18–25.

Mesic, Richard, David E. Thaler, David Ochmanek, and Leon Goodson, *Courses of Action for Enhancing U.S. Air Force "Irregular Warfare" Capabilities: A Functional Solutions Analysis,* Santa Monica, Calif.: RAND Corporation, 2010. As of February 15, 2010:
http://www.rand.org/pubs/monographs/MG913/

National Research Council, *Naval Analytical Capabilities: Improving Capabilities-Based Planning,* Washington, D.C.: National Academies Press, 2005.

———, *Defense Modeling, Simulation, and Analysis: Meeting the Challenge,* Washington, D.C.: National Academies Press, 2006.

———, *Conventional Prompt Global Strike Capability,* Washington, D.C.: National Academies Press, 2008.

NATO Command and Control Research Program (SAS-065), "NATO NEC C2 Maturity Model Overview (Draft)," 2009.

Pirnie, Bruce R., Alan J. Vick, Adam Grissom, Karl P. Mueller, and David T. Orletsky, *Beyond Close Air Support: Forging a New Air-Ground Partnership,* Santa Monica, Calif.: RAND Corporation, 2005. As of January 14, 2010:
http://www.rand.org/pubs/monographs/MG301/

Porter, Gene, David Berteau, Jay Mandelbaum, Richard Diehl, and Gary Christle, "Acquisition Initiatives Review—Phase II," briefing, 2006.

Porter, Gene, Jerome Bracken, Jay Mandelbaum, and R. Royce Kneece, *Portfolio Analysis in the Context of the Concept Decision Process,* Alexandria, Va.: Institute for Defense Analyses, 2008.

Simons, Robert L., *Levers of Organization Design: How Managers Use Accountability Systems for Greater Performance and Commitment,* Cambridge, Mass.: Harvard Business Press, 2005.

Smith, Edward A., *Complexity, Networking, and Effects-Based Approaches to Operations,* Washington, D.C.: DoD Command and Control Research Program (CCRP), 2006.

Snyder, Don, Patrick Mills, Adam C. Resnick, and Brent D. Fulton, *Assessing Capabilities and Risks in Air Force Programming: Framework, Metrics, and Methods,* Santa Monica, Calif.: RAND Corporation, 2009. As of January 14, 2010:
http://www.rand.org/pubs/monographs/MG815/

Third Infantry Division (Mechanized), "Close Air Support (CAS)," in *Third Infantry Division (Mechanized) After Action Report—Operation Iraqi Freedom,* StrategyWorld.com, 2009.

U.S. Air Force, "Air Force Basic Doctrine," briefing, November 17, 2003. As of January 14, 2010:
http://www.dtic.mil/doctrine/jel/service_pubs/afdd1.pdf

———, "Irregular Warfare: Air Force Doctrine Document 2-3," briefing, 2007.

U.S. Government Accountability Office, *Best Practices: An Integrated Portfolio Management Approach to Weapon System Investments Could Improve DoD's Acquisition Outcomes,* Washington, D.C., 2007.

Van Riper, Lt. Gen. Paul K. (USMC, retired), *Planning for and Applying Military Force: An Examination of Terms,* U.S. Army War College, Carlisle, Pa.: Strategic Studies Institute, 2006.

Vick, Alan J., Richard M. Moore, Bruce R. Pirnie, and John Stillion, *Aerospace Operations Against Elusive Ground Targets,* Santa Monica, Calif.: RAND Corporation, 2001. As of January 14, 2010:
http://www.rand.org/pubs/monograph_reports/MR1398/

Willis, Henry, James Bonomo, Paul K. Davis, and Richard Hillestad, *Capabilities Analysis Model for Missile Defense (CAM-MD): User's Guide,* Santa Monica, Calif.: RAND Corporation, TR-218-MDA, 2006. Limited distribution.